Microbes from Hell

Microbes from Hell

PATRICK FORTERRE

TRANSLATED BY
TERESA LAVENDER FAGAN

The University of Chicago Press
Chicago and London

The University of Chicago Press, Chicago 60637
The University of Chicago Press, Ltd., London
© 2016 by The University of Chicago
All rights reserved. Published 2016.
Printed in the United States of America

Originally published as *Microbes de l'enfer* by Patrick Forterre. © Editions Belin, 2007.

25 24 23 22 21 20 19 18 17 16   1 2 3 4 5

ISBN-13: 978-0-226-26582-7 (cloth)
ISBN-13: 978-0-226-26596-4 (e-book)
DOI: 10.7208/chicago/9780226265964.001.0001

Library of Congress Cataloging-in-Publication Data

Names: Forterre, Patrick, author. | Fagan, Teresa Lavender, translator.
Title: Microbes from hell / Patrick Forterre ; translated by Teresa Lavender Fagan.
Other titles: Microbes de l'enfer. English
Description: Chicago ; London : University of Chicago Press, 2016. | © 2016 | Includes
    bibliographical references and index.
Identifiers: LCCN 2016009697 | ISBN 9780226265827 (cloth : alkaline paper) |
    ISBN 9780226265964 (e-book)
Subjects: LCSH: Thermophilic microorganisms. | Microorganisms—Effect of heat on.
Classification: LCC QR84.8.F6713 2016 | DDC 579.3/17—dc23 LC record available at
    http://lccn.loc.gov/2016009697

Children of steam and scalded rock, a story you have to tell,
Writ in the glare of sunshine bright,
Sculptured and etched in marble white,
Illuminated in colors bold,
Richer than ever parchment old,
Children of steam and scalded rock, what is the story you have to tell?
Our legends are old, of greater age than the mountains round about.
We have kept our secrets epochs long,
They are not to be read by the passing throng.
It is nothing to us what men may say.
If they wish our story the price they must pay
In hard brain work, ere the tales are told. We challenge mankind to draw them out.
Children of steam and scalded rock, your challenge must rest for the present age.
I have scarcely broken the outer crust
That covers the greater truth, but I trust
Some man will follow and therein find
Knowledge, that to the Present shall bind
The Past with cords wherein entwine
Threads of the perfect truth, divine.
Children of steam and scalded rock, some man to come will accept thy gage.

A poem by B R A D L E Y   M O O R E   D A V I S, *University of Chicago, published*
in his 1897 paper in the journal Science *and describing for the first time*
microbes present in Yellowstone hot springs.

This book will tell you the story of those who "accept thy gage."

Contents

Prologue

The souls of the damned writhe in pain in the cauldrons of hell. Life appears to be incompatible with the flames of Satan. Plunge a crab into a pot of boiling water and it comes out cooked—but dead. Plunge your own finger in and the pain is unbearable. It's indisputable: even if we like heat, preferably in the shade on a beach, we must stay within the limits of what is reasonable. And yet, in the last thirty years, scientists have discovered microbes that thrive in temperatures that exceed those of the hottest deserts in the world. These creatures, whose existence is revealed only under a microscope, live in places that humans have equated with the gateway to hell. They are found in Solfatara, near Naples, Italy, which the ancient Greeks believed was the home of Charon, whose boat transported the souls of the dead on their final voyage, and in Iceland, the island where Jules Verne's heroes journeyed to the center of the earth. These microscopic single-celled organisms have telling names, such as *Acidianus infernus*, which likes both "infernal" temperatures and extreme acidity, or *Pyrococcus abyssi*, the "burning shell of the abyss." These are thermophile microbes (from the Greek *therme*, "heat," and *phylos*, "that which loves"). The most infernal among them have been baptized as "hyperthermophile" microbes because these amazing beings, which love the highest temperatures, flourish only between 80 and 110°C and freeze below 70°C.

These microbes from hell, whose existence was unknown to scientists until relatively recently, are today at the heart of particularly active research. International meetings have been devoted to them every two years since 1990, bringing together first dozens, then hundreds, of scientists from throughout the world. Why such passion for living beings so different from us? Some evolutionists believe that thermophile and hyperthermophile microbes can teach us a lot about the way life appeared on our planet, at a time when it perhaps resembled hell more than it did heaven. Others (sometimes the same ones) think the microbes that live in such extreme environments will help us determine the conditions necessary and sufficient for a planet to sustain life. Ultimately, all the specialists who study these microbes from hell wonder how they manage to thrive in places where all other forms of life are destroyed in a matter of seconds. The scientists' goal is to discover all the tricks these organisms have invented during evolution to protect their molecules from the destructive effects of extremely high temperatures. They tell themselves that these inhabitants of hot environments must be hiding many secrets that are only waiting to be brought to light by imaginative and persistent scientists, who can then make wonderful discoveries based on them.

The biotechnology industry and those who want to put microbes to work to replace traditional chemical products with more "green" chemicals are also very interested in these microbes from hell. They hope to be able to use the extraordinary resistance of the microbes' proteins for treatments that destroy "classic" proteins. These hopes are not utopian. One of the greatest technological revolutions known to biology in the past twenty-five years was accomplished thanks to an isolated protein of a thermophile microbe. This was the perfection in 1987 of a technique called PCR—polymerase chain reaction. This technique enables the controlled and specific amplification of any area of the DNA molecule: the famous double helix that contains our genetic data. This amplification alone enables us to obtain enough material to make the DNA molecule "speak." PCR has countless applications in biology, but also in medicine (in particular, for the diagnosis of hereditary diseases or viral infections) and even in criminology.

What is the connection between PCR and microbes from hell? The technique requires the use of an enzyme, DNA polymerase, capable of reproducing an identical DNA molecule by copying the two strands of the double helix (a chemical reaction corresponding to a polymerization). The amplification of DNA is achieved through the many cycles of chain reactions, during which the quantity of DNA is doubled each time. Each of these cycles includes an incubation stage at 90°C to separate the two strands of the double helix. The DNA polymerase used must thus be particularly strong at high temperatures. Here is the problem: at 90°C, the proteins are irrevocably destroyed—except, of course, those of microbes from hell.

The DNA polymerase used most often today to achieve the amplification of DNA through PCR is extracted from a thermophile bacterium called *Thermus aquaticus* ("hot microbe that lives in water"), discovered in 1967 by American microbiologist Thomas Brock in the hot springs of Yellowstone National Park. Without the curiosity and stubbornness of this scientist, many scientific discoveries would not have seen the light of day, many murderers would have escaped justice, and many innocents sentenced to death would have been executed.

The patents filed for the method that enables the amplification of DNA by PCR using the isolated enzyme of *Thermus aquaticus* and for the method that enables the amplification of DNA have for years brought in hundreds of millions of dollars every year. We now understand better why thermophiles and hyperthermophiles have been riding so high. For some, these microbes are veritable gold mines. In this work I will not, however, focus on this aspect of things, because a choice must be made, and I am not a specialist in biotechnology. Beyond being a gold mine for some and a source of innovation beneficial to all, microbes from hell are above all a passion for most of the scientists who are interested in them. I have therefore chosen to show why these living beings excite the curiosity of scientists and how this curiosity bears fruit. When Brock, while on vacation in Yellowstone, asked himself if the hot springs in the park might contain living beings, he couldn't have imagined that *Thermus aquaticus* would one day be a goose laying golden eggs. He was simply inspired by the desire to understand the world that surrounded and fascinated him. It was again curiosity that led two German biologists who will accompany us throughout this book, Wolfram Zillig and Karl Stetter, to travel the globe looking for creatures able to live at increasingly high temperatures. It is they who discovered hyperthermophiles. We will see how their research revealed a new world that until then had gone undetected, thus pushing the limits of the living being beyond anything we could have ever dreamed of before. They fulfilled the prediction made at the turn of the twentieth century in the poem (see the epigraph) by American biologist Bradley Moore Davis, who was the first to realize the existence of microbes from hell: "some man to come" would accept the challenge to understand these microbes, which Davis called by the poetic name "Children of steam and scalded rock."

Today, hyperthermophiles are no longer the only inhabitants of extreme conditions to be in the spotlight. All microorganisms that live in atypical physical and chemical conditions are in vogue. An increasing number of scientists are studying these microbes, known as extremophiles: some like high doses of salt (halophiles); others, intense cold (psychrophiles); others, very acidic or very basic environments (acidophiles and alkaliphiles); and so forth. However, in this work, I will focus only on thermophile and hyperthermophile microbes, leaving aside the other lovers of extremes.

My choice is in large part fairly personal. I've been studying the inhabitants of hot springs on Earth and in the sea for thirty years. But this choice is also understandable if one seeks to comprehend how the biosphere we know today was created. Indeed, if we find a large variety of living beings that like cold or salt, that resist high pressure or large doses of radioactivity (bacteria, but also algae, protists, insects, and fish), why have only two types of microbes—bacteria and archaea—managed to adapt to high temperatures? The answer to that question is probably rooted in the profound nature of cellular organization. Bacteria and archaea are distinguished from all other living beings by the absence of a nucleus enclosing the DNA molecule: they are prokaryotes (meaning "before the nucleus").

Everyone has heard of bacteria, but what are archaea (which for a long time were called archaebacteria)? In answering this question, I will have the opportunity to recount one of the most amazing upsets that has occurred in biology in the last thirty years: the American Carl Woese's discovery in 1977 of a new domain in the living world—a domain to which the most extreme of microbes from hell belong. The study of archaea has held many surprises for scientists, in particular for those interested in the archaea viruses. These scientists discovered many new viruses that were very different from the well-known viruses that infect bacteria and eukaryotes. Some of those viruses produce virions (viral particles) that change shape after leaving the cell; others produce pyramids capable of opening like flowers to allow their virions to escape from the cells they infect. Just as astonishing, these archaea have proven to be more closely related to humans (eukaryotes) than to bacteria. Many evolutionists today think that the archaea are in fact our ancestors. The debate on this issue, which directly involves our origins, has raged for some time. In 2013, I published an article titled "The common ancestor of archaea and eukarya was not an archaeon"; in my opinion, archaea are not our grandmothers but distant siblings . . . and bacteria are our cousins.

Then where do microbes from hell, in particular those archaea with their very strange name and their surprising viruses, come from? Did they adapt to "diabolical" conditions, or have they always lived with Lucifer? And how long have they been present on Earth? Did our own ancestors live in boiling water? Alongside these questions, the greatest debates have arisen, but also the most difficult and dangerous question that science has ever known: How did life appear on our planet? Indeed, some scientists today believe that life was born in hot springs: on the earth, under the sea, or underground. Suddenly, a seductive hypothesis was offered to them: the microbes from hell that we are studying today could be the direct descendants of the first forms of life that appeared on our planet more than three billion years ago. This could explain why the "hottest" of these microbes, those that live between 95 and 110°C, belong to the group of archaea ("the old ones")—a term that, when it was created, was meant to emphasize the assumed archaic side of its representatives. But we will see that other scientists, including me, have strongly contested this point of view. As with the question of our archaeon ancestor, the scientific community is divided. The arguments for and against are debated; scientific articles are written; discussions are . . . heated. In the last few years, several results seem to point to a new scenario that distances our microbes from hell from the origins of life, but still grants them an essential role in the history of the living being. I will leave you with the surprise of discovering this scenario in the work that follows. Before you do, though, we will learn more about the last common ancestor of all current living beings, known as LUCA (the last universal common ancestor), the very distant ancestor of the early human skeleton we call "Lucy." Did LUCA live in hell, or not? This is one of the questions I will attempt to answer.

Thus microbes from hell, whose existence was practically unknown just three decades ago, find themselves on center stage. What a spectacular turn of events! While describing these amazing organisms, I will attempt to respond to all the questions I have just raised. I will dissect them and make their macromolecules speak in order to discover some of their secrets, at times hidden in the heart of their genome. The answers to the questions raised will not be definitive; often they will lead to other questions, as always happens in research. We will see that many experiments remain to be done by future generations of scientists.

As a change of pace between two series of scientific explanations, I will also transport you to an encounter with microbes from hell at close to 3,000 meters under the sea, in the walls of hydrothermal chimneys deep in the ocean. We will even make a detour to Stockholm for the award ceremony of a prestigious science prize (not the Nobel). We will share a few convivial moments with scientists around a dinner table, paying tribute to French or Georgian gastronomy. And so bon voyage as we embark on a scientific adventure that has already provided a great deal of pleasure to those who have been lucky enough to participate in it, and which is not over yet.

A Bit of History: Microbes and Humans

Microbes from hell . . . Everyone has heard of microbes, and almost everyone has a vague idea of what the term represents: very small creatures that can do a lot of harm. If they're in hell, why not let them stay there? To get to the heart of the subject, we'll have to move beyond these prejudices, get to know microscopic life better, and look a little more closely at what the word "microbe" really means.

Did You Say "Microbe"?

For a long time, our knowledge of the living world was limited to macroscopic organisms—that is, those visible to the naked eye. The living world was generally divided into two groups: animals and plants. It was only in the seventeenth century, with the invention of the microscope, that the Dutch clothing manufacturer and microscope maker Antonie van Leeuwenhoek (1632–1723) discovered the existence of the then unimaginable world of microscopic living creatures. Two centuries later, Charles Sedillot, a contemporary and colleague of Louis Pasteur, would dub them "microbes," from the Greek *bios* for "life" and *mikros* for "small" (figure 1.1a).

Their sizes proved to vary considerably, with the largest among them measuring a few tenths of a millimeter in length and the smallest being no bigger than a micron (less than a thousandth of a millimeter). Very soon it appeared that these minuscule beings, invisible to the naked eye (or even the most powerful magnifying glass), were present everywhere: in water, in the air, in the ground, and even in the human body. In the nineteenth century, people were frightened to realize that these microbes, far from representing just a simple curiosity of nature that could be ignored, were the cause of many of

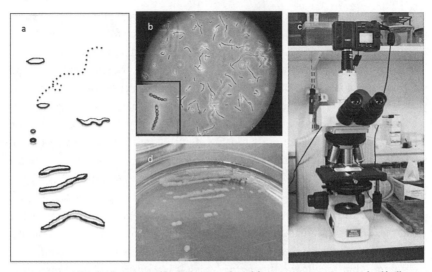

FIGURE 1.1. Microbes become visible. *A*, drawing adapted from Antonie van Leeuwenhoek's illustration of tiny microbes (most likely bacteria) using his homemade microscope. The dotted curved line reveals that these organisms were actively moving, an attribute of living beings. *B*, picture of bacterial filaments (*Streptococcus agalactiae*) observed with a modern light microscope, courtesy of Violette Da Cunha, Institut Pasteur. These filaments are strings of dividing bacteria (*b*, *inset*); each one is about one micron long (a thousandth of a millimeter). *C*, modern light microscope. *D*, petri dish with bacterial colonies visible to the naked eye on the surface of the agar layer. Each colony contains billions of bacteria that all originated via successive divisions from a single one that was placed on the dish.

the scourges that had stricken humanity for centuries: plague, typhus, tuberculosis, syphilis, and so on. Today we know that most microbes are innocent, and some are even our benefactors—in particular, those that live in our intestines and protect us from attacks by their pathogenic siblings (other microbes that are responsible for illnesses) and those that help us digest food. In the past few years, after sequencing the human genome, scientists have focused on the sequencing of the genomes of all microbes—the "microbiome." In fact, we contain ten times more microbial cells than human cells; these microbes belong to a considerable number—several thousand—of different species. Scientists have recently discovered that these microbes might even manipulate the brain: for example, by producing molecules that make us want to eat what they need. If we get to know them better and learn to manipulate them in turn, we should be able to derive enormous benefits in the realm of human health. Nevertheless, the term "microbe" in the collective subconscious retains a rather fearsome connotation. (I hope that after reading this book you will have changed your mind a bit.)

In general, biologists no longer use the word "microbe" (perhaps due to its rather pejorative nature, or because it doesn't sound very scientific). Instead, they use the word "microorganisms" to describe the living beings that can be observed only under a microscope. For a long time, scientists were unable to situate these organisms in relation to other representatives of the living world. They even tried to force them into the old traditional framework by seeking characteristics in their appearances that might connect them to animals or plants. If a microbe was able to move on its own, thanks to a flagellum, it was a minianimal, and it became a subject of study for zoologists. If, on the contrary, it remained immobile most of the time and was made up of a cell surrounded by a thick wall, it was a miniplant, and was of concern to botanists. It was from that time that biologists, in order to name microbes, began to use binomial nomenclature (in which every living being is designated by a genus name and a species name), invented in the seventeenth century by the great Swedish naturalist Carl von Linné (Linnaeus) for animals and plants. For example, the bacterium responsible for plague was named *Yersinia pestis*, a name that recalls the illness for which it is responsible and honors the one who discovered it: Pasteur's young colleague Alexander Yersin. The name of the bacterium used as a model by many biologists, *Escherichia coli* (commonly called *E. coli*), recalls its preferred habitat, our colon, and honors another microbiologist, Theodor Escherich, who was working in Vienna at the beginning of the twentieth century.

A decisive tool in the study of microbes, the petri dish (figure 1.1d), was invented in the 1880s in Germany by the team of Pasteur's German rival, Robert Koch, who discovered *Mycobacterium tuberculosis*, responsible for tuberculosis. The story goes that it was while observing colonies of bacteria and mushrooms on the surface of slices of potatoes, the leftovers of a meal eaten quickly between two experiments, that the German scientists understood that each colony (circular spots of varying size, of different colors, whose edges were more or less sharp) corresponded to a clone of billions of microbes, all of which came from a single cell, bacterium or mushroom, that had settled on the surface of the potato. Julius Richard Petri, one of Koch's assistants, had the idea of filling a circular glass dish with a jellylike substance, agar, containing a culture medium to obtain bacterial colonies isolated on the surface of the agar (figure 1.1d). He had only to dilute the bacterial cultures sufficiently to place a few bacteria on the petri dish. This method, still used today, makes it very easy to isolate pure cultures of bacteria, which can then be characterized; this enables species to be defined. The glass of Petri's dishes has simply been replaced with plastic; agar, initially used by the wife of another of Koch's assistants to cover the surface of her jam jars to keep them sterile, is also used today.

Microbes Are Cells

During the nineteenth century, scientists gradually came to understand that all living beings, which at the time they classified in the category of either animals or plants, were formed by the assembly of a very great number of cells (a hundred thousand billion for a human being), with each cell forming the morphological unit—the elementary brick, in a sense—of the living parts of organisms. Observed under an optical microscope, the cells appeared as masses of a gelatinous substance (cytoplasm) surrounded by a thin membrane and containing a spherical central region enclosing nucleic acids (observed by using certain dyes): a sort of nucleus. Animals, plants, most algae, and mushrooms were thus formed by a great variety of organs and tissue, with each tissue composed of myriad cells. It was then noted that most microbes were formed by a single cell. Alongside the division of the living world into animals and plants, a division between multicellular (what I sometimes call macrobes) and single-celled organisms (microbes) came to be superimposed.

It quickly became obvious that single-celled organisms had an incredible variety of shapes: there was a microbe, and then there was a *microbe*. The largest could resemble animal or plant cells that had been able to live in an isolated state. This was, for example, the case of the paramecium, which, with its single cell endowed with two orifices—one serving as a "mouth," the other as an "anus"—seemed to sum up the functions of a complete animal organism. Such living beings, which are grouped under the term "protists," were for a long time considered to be small, primitive animals or plants—protozoa and protophyta, respectively. Other microbes, such as yeasts, presented characteristics common to mushrooms, while still others were similar to algae. Alongside these miniature animals or plants were even smaller microbes, already observed by Van Leeuwenhoek, that could not be connected to any group of organisms visible to the naked eye. These "minimicrobes," although visible under the optical microscope, were so small that using traditional optical microscopes to distinguish any details inside the single cell that formed them was impossible (figure 1.1b). In particular, it was difficult to distinguish a nucleus in these miniature microbes. The most characteristic among them had the shape of a stick—hence their name, "bacterium," from the Greek term *bakteria*, which means "stick." Subsequently, all extremely small microorganisms formed by a cell of this type were called bacteria, regardless of their shape: stick, tendril, sphere, or filament. Until the end of the nineteenth century, botanists associated bacteria with plants, owing to their thick cellular wall. During my studies at the Faculté des Sciences in Paris, in 1969, I heard of

bacteria for the first time in a course in plant biology—proof that in matters
of classification of living beings, ideas can take time to evolve. We will soon
see that bacterial cells are quite different from plant cells.

Enter the Viruses

Toward the end of the nineteenth century, a new class of microbes made a
surprise appearance: viruses (a Latin word that means "poison"). These microbes, responsible for infectious illnesses, were so small that they could not
be observed under optical microscopes. Then how were they detected? It was
noticed that, even after filtration through the holes of porcelain filters capable
of retaining the smallest bacteria known at the time, the secretions of plants
or animals afflicted with certain illnesses (tobacco mosaic or rabies, for example) still contained a microbe capable of infecting another organism. In
addition to their small size, viruses presented another specificity: they could
not be cultured alone, unlike most of the bacteria studied in laboratories.
They could reproduce only in the presence of the organism they infected.
Viruses were thus obligatory parasites, but scientists didn't understand why.

At the beginning of the twentieth century, a French Canadian scientist,
Felix d'Herelle, discovered viruses that attacked bacteria. He gave them the
name "bacteriophages," meaning "eaters of bacteria," because at the time most
of his colleagues didn't believe they were viruses. D'Herelle immediately had
the idea of using these viruses to treat illnesses caused by pathogenic bacteria.
A few days after discovering bacteriophages (often simply called phages) in
the stool of a patient hospitalized at the Institut Pasteur who was beginning
to feel better, d'Herelle was able to cure a little girl of typhoid. This was the
beginning of phage therapy, a method of treating bacterial illnesses by using
phages. This therapy was commonly used in the Soviet Union, but rarely in
the West, in particular after the discovery of antibiotics. Today, phage therapy is coming back into fashion, owing to the growing number of dangerous pathogenic bacteria that have become resistant to multiple antibiotics.[1]
Petri's cultures, mentioned above, also played an important role in the study
of phages. If you grow a film of bacteria on the surface of a petri dish (this is
called a lawn) and you place a few phages on the surface of this film, you will
see light spots appear. These correspond to the places where, in multiplying,
the phage has gradually destroyed all the bacteria (figure 1.2e).

This is called a lysis plaque, because the bacteria have been "lysed" by the
virus, meaning their cell structure has burst. This technique has, once again,
enabled the definition of viral species through the pure viral cultures that
have been obtained. All the same, faced with uncertainty regarding the nature

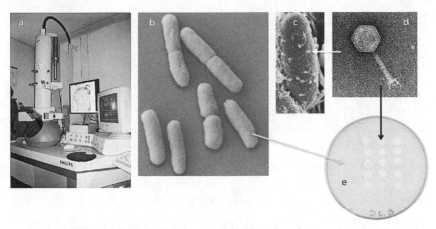

FIGURE 1.2. Viruses infecting bacteria become visible. *A*, modern electron microscope. *B*, scanning electron micrograph (SEM) of the rod-shaped bacterium *Escherichia coli* (about two microns long), with three bacteria in the process of cell division. *C*, SEM of a bacterium with virions attached to its surface (*white arrow*). *D*, transmission electron micrograph (TEM) of a head and tailed virion: the length is about 0.2 microns, a twenty-thousandth of a millimeter! *E*, a petri dish covered with a layer of bacteria (*gray arrow*) on which fifteen virions have been placed (*black arrow*). These virions initiated a viral infection that spread from bacteria to bacteria, disrupting their bodies. This led to the formation of fifteen holes (plaques) in the bacterial layer. Courtesy of Laurent Debarbieux, Institut Pasteur.

of these microscopic parasites, the Linnaean binomial nomenclature was not used for viruses; instead, they received various names, most of the time associated with the illness for which they were responsible (such as the AIDS virus). Bacteriophages, which are all responsible for the same "illness" (the death of bacteria), are sometimes given names associated with the infected bacteria, or, when a bacterium was infected by different phages, are named through associations of numbers or letters (or both) that recall the conditions in which they had been isolated. We will see a few examples of this below.

When Two Primary Kingdoms Share the World of the Living

For a long time, many biologists had a tendency to consider bacteria and viruses as groups completely separate from the rest of the living world. They were both right and wrong. We had to wait until the middle of the twentieth century to realize that these microbes were composed of the same molecules as animals, plants, or mushrooms. In the 1960s, the pioneers of molecular biology were able to show that the principal processes ruling the functioning and division of a cell are the same in all cellular living beings: bacteria, protists, animals, or plants. It was at that time that the French scientist Jacques Monod, who won the Nobel Prize in Medicine in 1965 for his discovery of the regulatory mechanisms for the expression of genes, coined a famous phrase:

"What is true for *Escherichia coli* [the model bacterium for molecular biologists] is true for the elephant." Let's briefly recall that in all living beings, the hereditary genetic data are contained, in the form of genes, in a DNA molecule (or RNA, in certain viruses). Each gene encodes the message enabling the synthesis of a protein on the level of a specialized cellular structure: the ribosome. Each protein serves a given function in the life of the cell. The genetic code, which enables the translation of the message carried by the DNA in proteins, is identical for all living beings. We will return in more detail to all these mechanisms in chapter 3.

All the same, this unity of the living world must not hide the fact that there are profound differences within it. When biologists were able to use electron microscopes (figure 1.2a) to observe microbes, they realized that the bacterial cell was much simpler than the cells of all other living beings. The cells of animals, plants, mushrooms, or large microbes (protists) are particularly complex. They contain all the specialized "cell organelles" (such as mitochondria, which are used in breathing). Observation under the electron microscope also revealed an elaborate system of intracellular membranes inside these cells. In particular, their chromosomes, which contain the genetic information in the DNA molecule, are separated from the cytoplasm by a membrane, the nuclear membrane, that delimits a specific compartment, the nucleus, which had already been observed under an optical microscope (see the two pictures of an amoeba in figure 1.3).

This compartmentalization has a very important effect: proteins are synthesized on the ribosomes located in the cytoplasm. The genetic message, contained in the nucleus, must then be transferred from the nucleus to the cytoplasm to be able to be read by the ribosomes. Molecular biologists have shown that each gene sends ribosomes a "messenger" RNA that is synthesized in the nucleus and then transported to the cytoplasm by passing through the pores located in the nuclear membrane (figure 1.3).

Bacterial cells are much simpler (and often much smaller: see figure 1.3 for an example). In general, they don't contain any organelle or intracellular membrane (as always in biology, there are exceptions[2]), and they don't possess a "true" nucleus: their genetic material is localized in the cytoplasm, directly in contact with the machinery that creates the proteins. This proximity enables the RNA message to be read by the ribosomes in the process of its synthesis through bacterial DNA, called a coupling between the transcription (transcription of the DNA message into a messenger RNA) and translation (translation of the message into proteins; figure 1.3). We will return to these mechanisms in more detail in chapter 3.

The complex cells of animals, plants, and protists have been called eu-

Prokaryote (ARCHAEA, BACTERIA)

Eukaryote (EUKARYA)

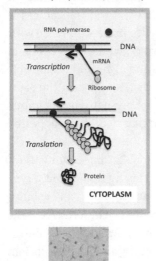

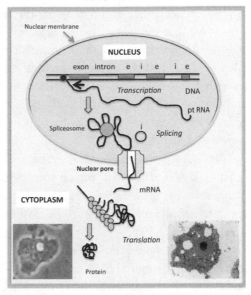

FIGURE 1.3. Prokaryotic versus eukaryotic cells. In prokaryotes (archaea and bacteria), the transcription of DNA into messenger RNA (mRNA) is coupled with the translation of the mRNA into proteins by the ribosomes. In eukaryotes, these processes are uncoupled, because transcription occurs in the nucleus, whereas translation occurs in the cytoplasm. Furthermore, eukaryotic genes are split into coding sequences, called exons (e), and noncoding sequences, called introns (i; discussed in chapter 2). The entire genes are first transcribed to produce primary transcripts (ptRNA). These ptRNA are spliced by the spliceosome inside the nucleus to form the final mRNA. Splicing of the ptRNA is coupled with the transfer of the mRNA to the cytoplasm via nuclear pores. The optical micrograph below the prokaryotic cell represents a mixture of bacteria (bacilliform) and archaea (cocci) (courtesy of Évelyne Marguet, Université Paris-Saclay). The optical micrograph and TEM (*from left to right*) below the eukaryotic cell represent an amoeba (courtesy of Morgan Gaia, Institut Pasteur). Large white vacuoles can be seen in both pictures, whereas the nucleus (*black sphere*) can be best observed in the TEM picture. All pictures are roughly of the same scale.

karyotic (which literally means "true nucleus") to distinguish them from bacterial cells, including DNA, which is simply condensed at the center of the cell, forming a "pseudo" nucleus, explaining why it was difficult to observe under an optical microscope. These cells without a true nucleus have been called prokaryotic, which can mean "approximate nucleus," or, more literally, "before the nucleus."

The discovery of these two cell types within the biosphere led biologists in the 1960s to divide the living world into two large "superkingdoms": eukaryotes, including animals, plants, mushrooms, algae, and protists; and prokaryotes, containing only bacteria. This classification was quickly adopted by all biologists; very soon it was also considered to be a reflection of the history of the living world. Prokaryotes, being simpler, must have appeared first, followed much later by eukaryotes. The term "prokaryote," meaning "before the nucleus," reflects that idea. Following this hypothesis, it was noticed at the same time that the two principal types of organelles present in eukaryotic cells—mitochondria (present in almost all eukaryotes, with the exception of a few protists) and chloroplasts (present in plants and algae and responsible for photosynthesis)—came from ancient bacteria that, a very long time ago, had merged through symbiosis with the ancestors of present-day eukaryotic cells. The modern eukaryotes (with mitochondria) thus necessarily appeared after bacteria.

Very quickly, the terms "prokaryote" and "bacteria" became synonyms for most biologists. The term "prokaryote," which originally designated a type of cell structure (absence of a true nucleus), ended up taking on an evolutive connotation: all prokaryotes must be close relatives, and all descend from the same ancestral prokaryote. For many biologists, the molecular mechanisms of prokaryotes should thus greatly resemble each other, and it was sufficient to study one bacterium on the molecular level to know them all. In particular, the discoveries made from the model bacterium for molecular biologists, *Escherichia coli*, were to be systematically extrapolated to all bacteria. In fact, this way of thinking, still tacitly accepted today by many biologists and taught in most universities, would be proven to be completely false, as revealed at the end of the 1970s in some truly revolutionary work. We will soon discover that microbes from hell played a primary role in this affair.

Before getting to that, let's go back to viruses, those mysterious obligatory parasites. For a long time, biologists thought that viruses were nothing more than miniature bacteria, too small to be seen under an optical microscope. The possibility of seeing them, thanks to the electron microscope, was a revelation. They no longer resembled bacteria at all, or even a cell, prokaryotic or eukaryotic. We have all seen viruses represented on television in the form of an evil-looking thorny sphere. In fact, viral particles, called virions, appear in a great variety of shapes. (We will see below why I make this distinction between viruses and viral particles.) Many viruses indeed produce spherical virions, of sometimes geometrical structures (icosahedrons), but others, such as Ebola, produce filament-like particles or particles in the shape of rigid sticks, such as the tobacco mosaic virus. Some virions are quite curious, such as the one represented in figure 1.2d, which infects bacteria. This virion, typical of many bacteriophages (borrowing d'Herelle's terminology for bacterial viruses), has a "head," a "tail," and "feet," resembling a small animal.

Viral particles have a point in common: they don't resemble cells at all (compare in figure 1.2 the bacterium *Escherichia coli* and the virion of a bacteriophage infecting this bacterium, observed by electron microscopy). Virions don't have cytoplasm and, with some exceptions, they are not surrounded by a lipid membrane. They are essentially made up of a more or less complex shell (spherical virions) or cylinder (filament virions), called a capsid, formed by the grouping of many proteins. This capsid contains the virus's genetic material in the form of DNA or another molecule, RNA, which we'll talk about again. An essential point: the virion does not have the enzymes necessary for metabolic functions (assimilate food and/or chemical or luminous energy and produce the energy to be used for biological functions). Similarly, the virion does not possess the machinery (ribosomes) that enables the synthesis of proteins from its genetic information. Its genome doesn't contain the genes necessary for manufacturing a ribosome. To synthesize their proteins and multiply, viruses must thus be able to use the ribosomes of a cell, which explains why they are indeed obligatory parasites that can reproduce only by infecting cells, whose ribosomes they use. For that purpose, they must find a way for their genetic material (DNA or RNA) to penetrate a prokaryotic or eukaryotic cell. In figure 1.2c, you can see many viral particles attached to the surface of a bacterium. One of them will succeed in injecting its DNA into the cytoplasm of the bacterium by perforating its cell membrane. Once there, the viral message will usually bypass the cellular one, forcing the cell to synthesize viral proteins instead of cellular proteins.

Viruses could thus not be easily classified within the living world; they are neither prokaryotes nor eukaryotes. To borrow the expression of a great specialist of viruses from the previous century, the Nobel laureate André Lwoff, viruses are . . . viruses. They are unique microbes (some biologists even refuse to rank them in the category of microbes). We will see that viruses from hell are some of the most interesting ones.

The Discovery of the First Microbes from Hell

The "official" discovery of the first microbes from hell occurred shortly after it was proposed that the living world be divided into eukaryotes and prokaryotes, in the 1960s. Why did it take so long for the existence of these creatures to come to our attention? To answer this question we should ask why it should be surprising to find living beings in hot springs if, as we often say, microbes are everywhere.

WHEN HEAT DESTROYS LIFE

The work done in the nineteenth century, in particular by Louis Pasteur (1822–95) in France and Robert Koch (1843–1910) in Germany, showed that bacteria and viruses play a determinant role in many human illnesses. The development of hygiene practices at the end of the nineteenth century coincided with this awareness: our hands can transport pathogenic viruses or bacteria from one patient to another, our food could be contaminated, and so on. One of the major concerns was thus to understand how one could get rid of those microbes. Very quickly it was seen that a large majority of them were sensitive to detergents, to heat, or to alcohol. Sterilization by heat (such as pasteurization, which occurs at a temperature close to 80°C) became the norm for preparing many different foods. High heat thus seems to be the enemy of bacteria and viruses, as well as of the rest of the living world. Of course, there are bacteria that in difficult conditions can transform themselves into spores capable of surviving at very high temperatures. But it's just a matter of survival: a spore doesn't divide; it is completely passive in the face of the outside world. In the spore, the dormant bacterium patiently waits for the external temperature to become livable again to come out of its "shell." Even prions, those particularly resistant infectious proteins, can be deactivated if they are heated long enough and at a high enough temperature (we could thus have avoided mad cow disease if, for economic reasons, certain corporations involved in the manufacture of animal flour had not decided to reduce their heating costs).

WHAT IS TEMPERATURE?

To decontaminate, then, you must heat. Indeed, the forms of life that we are used to encountering around us do not survive temperatures that are too high. But what do we mean by "temperatures that are too high"? We know that temperature is associated with the living through subtle connections. For humans, the right temperature is 37°C. Above that, one has a fever, and below it, hypothermia. In both cases, if one goes beyond very precise limits, death can occur. To answer our question, we must begin by asking, "What is temperature?" A physicist might respond, "Temperature is a measurement of the agitation of atoms."

That statement plunges us into the heart of the matter. Everyone knows that atoms are the elemental bricks from which all objects present in our universe are formed (our bodies, the planets and stars, clouds in the sky, the water of the oceans, and so on). Atoms can exist in a free state or be connected to form molecules of extremely diverse natures. And so, matter is in perpetual motion: atoms and molecules move. The hotter it is, the greater the agitation: −100°, −10°, 0°, 10°, 100°, 1000°C. Temperature can thus reach colossal numbers (15 million degrees at the heart of the sun, for example).

Inversely, one can also calculate the temperature at which atoms and molecules remain completely immobile: this is absolute zero, which corresponds to −273°C.

WHAT IS THE "RIGHT TEMPERATURE"?

When they move, atoms collide with each other. These collisions enable matter to be formed. In particular, the collisions cause the formation of molecules by facilitating the combination of atoms. Agitation, and thus a certain temperature, is therefore indispensable for matter to take shape. But too much agitation has the inverse effect: molecules then "burst," in a sense, and complex structures cannot exist anymore.

Living matter is, beyond a shadow of a doubt, the most elaborate form of organization of matter that we know of for the time being. In particular, all cells, tissues, and other organs are built from giant molecules called macromolecules, formed from thousands or even millions of atoms positioned very precisely in space. These macromolecules are present only in living beings and are much more complex than all the other molecules in the universe that are currently known. This complexity of the living also explains its fragility. Have you ever experienced this while looking at mountains? They will still be there in millions of years, whereas our eyes that observe them are on Earth for less than a century. But living matter remains particularly sensitive to temperature: if it rises too much, if the agitation of atoms goes beyond a certain threshold, all that wonderful organization at the very foundation of what we call "life" risks collapsing like a house of cards.

Let's look a little closer at the macromolecules of living beings to illustrate this point. Let's look at what happens, for example, when an egg is cooked. An egg is made up of a shell, the white, and the yolk. Let's break the shell. The white and the yolk run out in the form of a nutritious liquid, the one in which the future little chick was swimming. That liquid is essentially made up of a specific type of macromolecule unique to living beings: proteins (figure 1.4c, d).

A certain number of these proteins (e.g., those of the white and the yolk) have the shape of little irregular globules of a few thousandths of microns in diameter. Each sphere (thus each protein) is formed by a long filament—a pearl necklace, if you will—folded on itself, like a little ball of string (figure 1.4a). This folding is the result of the interactions among the pearls, which, as we'll see below, correspond to amino acids (figure 1.4b). When you cook an egg, the agitation of atoms on the burning surface of the pan is transferred to the amino acids of the proteins of the white and the yolk, the pearls are agitated, and the balls unwind. The proteins lose their spherical appearance

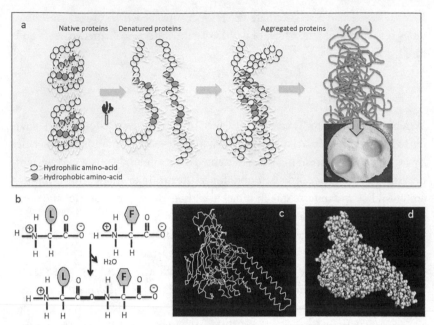

FIGURE 1.4. Denaturation of proteins at high temperature. *A*, the polypeptide chain is folded onto itself by the formation of weak bonds (*dotted lines*) between distant amino acids (*circles*) (see chapter 2). These weak bonds are eliminated at high temperature, leading to the denaturation of the proteins. The unfolded protein chains then aggregate via the formation of weak bonds between hydrophobic amino acids (*gray circles*). This leads to the formation of a network of denatured proteins that can include billions of entangled polypeptide chains, as in the case of fried eggs. *B*, formation of a peptide bond (*thick lines*) between two amino acids, L and F. The gray circles and icosahedrons symbolize atom groups that are different. *C, D*, two representations of a real protein: human type I DNA topoisomerase (see chapter 2 for an explanation of DNA topoisomerases). These models (courtesy of Daniele Gadelle, Université Paris-Saclay) represent schematically either the path of the polypeptide chain in the protein (*C*) or all atoms present in the protein (*D*).

because the pearl necklace of the amino acids unfolds and agitates randomly as it encounters the water molecules that also make up the egg: biochemists say that the proteins are denatured. Once unfolded, the necklaces of the different protein molecules very quickly attach to each other and assemble into a shapeless mass: they have precipitated (figure 1.4a). Instead of macromolecules isolated from each other, we now have a mixture of necklaces tangled into a giant three-dimensional network that can no longer "move." In the pan, the white of the egg has coagulated, the yolk begins to harden, and the fried egg is ready for you to enjoy.

The denaturing of proteins is one of the major consequences of a rise in temperature. If the proteins precipitate, the process is irreversible. When a living being is placed in an environment where the temperature is too high, such a phenomenon is produced en masse, and it means certain death. Let's take a classic microbe, the *Escherichia coli* bacterium, which lives peacefully in our intestines at 37°C. Plunge it into boiling water and in an instant it loses its beautiful stick shape, retracts, and dies. All its proteins have denatured. Stick your finger in boiling water. . . . No, that's a joke, don't do it. Oh, too late. . . . Look at that burn on your finger: the skin has cracked—the proteins of your epidermis have denatured. Biochemists also know that the proteins they study are fragile macromolecules that are sensitive to heat: today, to purify them, they use refrigerated coolers to keep them from denaturing. When I purified proteins myself, while writing my PhD thesis, we usually worked in a cold chamber (at 4°C). (I remember a hot summer day when I had come to work in the lab in shorts. I had completely forgotten that I was supposed to isolate a protein that day, and thus was to work in the cold chamber. Luckily, the lab was located in the heart of Paris, and I was able to run out and buy a pair of pants!)

Let's return now to the original question, that of the "right temperature." The few experiments that I've just described seem to indicate that life can exist only at temperatures lower than those at which proteins (and other macromolecules of the living) are denatured. And yet . . .

YELLOWSTONE: A BOILING PARK

How could a biochemist who takes so many precautions to avoid the denaturing of his favorite proteins ever imagine microbes living at 100°C? For a long time, the very idea that bacteria, viruses, or protists could live in hell never crossed the minds of most biologists. In fact, they would have declared that the very existence of such microbes was impossible. But one must beware of certitudes. Two English mechanical engineers at the beginning of the nineteenth century, Richard Trevithick and Andrew Vivian, declared that a locomotive with iron wheels would slip off iron rails; this belief caused a delay of ten years in railroad research, before a third man, Christopher Blackett, who was more adventurous, tried the experiment and contradicted the theory.

In the story that concerns us, there have been a few precursors.[3] In 1897 an American botanist, Bradley Moore Davis, published in the journal *Science* a description of the "vegetation" of the hot springs of Yellowstone National Park. (Indeed, we have seen that until the end of the nineteenth century, botanists often assimilated bacteria with plants.) He described the presence of microbes looking like bacteria in springs at temperatures up to 80°C and suggested that the upper temperature for life could be between 80 and 85°C. In 1903 another American biologist, William Setchell, also observed in Yellowstone bacteria that were active at high temperatures of 89°C. He had already

FIGURE 1.5. *A*, discovering amazing inhabitants of Yellowstone hot springs. *B*, an optical micrograph of the bacterium *Rhodothermus*, a relative of the first thermophilic bacteria isolated from Yellowstone hot springs, *Thermus aquaticus*. Courtesy of Claire Geslin, Université de Bretagne occidentale. *C*, SEM of the archaeon *Sulfolobus*, the first hyperthermophilic archaeon isolated from Yellowstone, courtesy of David Prangishvili, Institut Pasteur. *D*, a student's view of *Sulfolobus*, an extremophile growing at 83°C and pH2 (author's collection). *E*, a liquid culture of *Sulfolobus* made in 1983 in the author's laboratory using a magnetic stirrer with heat (see chapter 2).

raised this question: "What is it that enables the protoplasm [today we say cytoplasm] of the thermal organisms to withstand a temperature which coagulates, and consequently kills, the protoplasm of the majority of organisms?" He reckoned that "there may be some important difference in the essential proteins of the mixture or in the nature of the substance (the cytoplasm) which renders it less coagulable." However, most biologists ignored the work of these pioneers for many years.

From the very beginning, the hot springs of Yellowstone (figure 1.5a) have thus played a considerable role in the history of research on microbes from hell. Yellowstone owes its fame to the many constantly active geysers; to its hot springs painted in green, brown, yellow, and red by the myriad microorganisms in them; and to its yellow canyon carved out of a plateau of rocks rich in sulfur (whence the name "Yellowstone"). It is one of America's favorite tourist destinations. In the 1950s, one of those vacationers must have decided to make it his workplace. This visionary was Thomas Brock, a microbiologist. He is considered today to be the father of microbes living at high temperatures, known as thermophiles, to which would later be added microbes that are even more extreme—hyperthermophiles, which we will soon discover.

Of course, as we have just seen, scientists before Brock knew of thermophile microbes, but they had remained almost unnoticed. The observations made by Davis and Setchell at the turn of the 1900s on the "vegetation" of Yellowstone had been forgotten. The most thermophile microbe widely studied at the time was *Bacillus stearothermophilus*, whose optimum temperature—the temperature at which the microbe multiplies most efficiently in the laboratory—is 55°C. Already, very few biochemists would have dared to expose their favorite protein to that temperature. Seeking organisms capable of enduring higher temperatures appeared truly incongruous. Brock was to shatter that prejudice by confirming the observations of Davis and Setchell and by isolating the microbes that those scientists had glimpsed only by chance. The best known of these "children" of Brock, which are mentioned in the prologue, is *Thermus aquaticus*. Discovered in 1967, its maximum temperature for growth—the maximum temperature at which, in the laboratory, the microbe can still divide—is 78°C[4] (see figure 1.5b for a close relative of this bug). In 1972 Brock succeeded in isolating a new thermophile microbe in another hot spring of Yellowstone, this one capable of dividing at up to 85°C: *Sulfolobus acidocaldarius* (its name referred to its morphology—a lobed cell—and its environment—a hot and acidic environment containing sulfur; figure 1.5c).[5] In discovering this "extreme" thermophile, Brock had, without realizing it, opened a new chapter in the history of science.

THE APPEARANCE OF A THIRD "PRIMARY KINGDOM" IN THE LIVING WORLD: THE ARCHAEA

The discovery of *Sulfolobus* would indeed mark a turning point in the study of microbes from hell. This is where the American molecular biologist Carl Woese comes onto the stage. Woese was a physicist by training who had never isolated a microbe. At the time, he wasn't really interested in thermophiles, but in evolution with a capital E. What he wanted was nothing less than to retrace the history of the entire living world: to construct a "genealogical" tree of the entire earthly family, from bacteria to man—the universal tree of life. At the beginning of the 1970s, we saw that the only known organisms with prokaryotic cells were bacteria. To express this schematically, one might say that the history of life on our planet was envisioned as a two-stage rocket.

The first stage, at the base of the universal tree of life, corresponded to the diversification of prokaryotes from a primitive ancestor, itself of bacterial type. In the second stage, we witnessed the evolution of eukaryotes from an ancestor formed by the association of several bacterial cells. This vision of the history of the living did not, however, rest on any scientific data. Bacteria were simply placed at the base of the universal tree because of their small size and the relative simplicity of the prokaryote cell compared to the eukaryote cell. The very etymology of the word "prokaryote"—"before the nucleus"— seemed to imply that eukaryotes appeared after prokaryotes. We find here a vision of the evolution of the living world that was not unlike Aristotle's famous Great Chain of Being.

To go beyond appearances, Woese undertook the construction of the universal tree of life on a solid scientific foundation. To do this, in collaboration with his colleagues at the University of Illinois at Urbana, he sought to determine the kinship relationships between various organisms in the living world by using a method unknown up to then. In determining kinship relationships, one asks, "Who is the closest relative to whom?" and not "Who descends from whom?" The latter question is forever without an answer, as there are no civil records of the living world. To set forth this difference, scientists do not speak of a genealogical tree, but of a phylogenetic tree (from *phulon*, which means "tribe," Latinized into *phylum*, and *genos*, "origin").

To construct such a tree, Woese chose to study a macromolecule present in all living beings, from bacteria to man, called 16S ribosomal RNA (we will see in chapter 3 what 16S means). It is one of the components of ribosomes. This macromolecule is thus universal—in other words, it is found in all cells—since all cellular living beings need ribosomes to manufacture their proteins. We will see in chapter 3 what RNA is and how ribosomal RNA can be compared in organisms as different as a bacterium, a yeast, a plant, and an animal. For the moment, it's enough to know that the analyses carried out by Woese and his team were based on the principle of "that which resembles assembles." In other words, the scientists compared different ribosomal RNA (rRNA) molecules—which in some way played the role of genetic imprints— and the more the rRNA of two organisms resembled each other, the more closely related the organisms should be.

Initially, this new approach enabled the scientists on Woese's team to find the division between prokaryotes and eukaryotes detected thirty years earlier by cellular biologists with the help of the electron microscope. They were then able in 1975 to show that the chloroplasts present in green plants indeed came from former bacteria.[6] The great surprise came in 1977, when Woese became interested in microbes that had rarely been studied up to then: the "methanogenic bacteria," or producers of methane. These microbes are capable of causing a molecule of carbon gas (CO_2) and four hydrogen molecules

(H_2) to react to create two water molecules (H_2O) and one methane molecule
(CH_4). This latter molecule is an inflammable gas; the presence of methanogenic organisms at the bottom of certain swamps explains the appearance of
will-o'-the-wisps on their surface, when the methane produced in the depths
rises to the open air, where it encounters oxygen molecules. Methanogens live
in oxygen-free environments, such as lake or marine sediments, certain hot
springs, rice paddies, and even certain zones of our intestines: these are called
anaerobic organisms.

To the great surprise of Woese and his colleague, George Fox, although
the cells of the methanogens were of a prokaryotic type (they don't have a
nucleus), their genetic imprint proved to be as far from that of classic bacteria
as from that of humans and other eukaryotes. Woese's heart skipped a beat;
without hesitation, he decided to leave his name to posterity by proposing to
give up the old binary division of the living world into eukaryotes and prokaryotes and to replace it with a division into three great "primary kingdoms"
(*urkingdoms*): eukaryotes, classic bacteria, and a third group corresponding
to methanogenic bacteria.[7]

This was no small matter, because people really like binary divisions:
right/left, more/less, yin/yang, male/female, and so on. To drive the point
home, Woese decided to give the new grouping a catchy name. He chose
"archaebacteria"—literally, "old bacteria"—stressing the anaerobic nature
of methanogenic bacteria and their demand for an atmosphere rich in carbon gas and hydrogen. We know for certain that the atmosphere of primitive
Earth, around four billion years ago, did not contain oxygen, and it is believed
that it was rich in carbon gas, the result of the degassing associated with the
formation of Earth on the one hand, and eruptions by volcanoes, quite active
at the time, on the other. It was thus tempting to think that methanogenic
microbes were able to populate such an environment. At the same time as
Woese proposed the new name of archaebacteria for methanogenic bacteria,
he proposed the name "eubacteria" for all other bacteria.

The concept of archaebacteria revolutionized biology by raising many new
questions. We'll return to them in chapter 4, but let's ask one of the most
important of them now: What did the common ancestor of the three lineages thus presented look like? This was a virtual organism that Woese called
the progenote and that today we call LUCA (last common universal ancestor).[8] For the linear vision that ruled up to then (from primitive bacteria to
evolved eukaryotes), a much more complex history was substituted. Instead
of the two-stage rockets of the traditional scenario, the new vision that Woese
proposed suggested three lineages and a mysterious ancestor (figure 1.6).

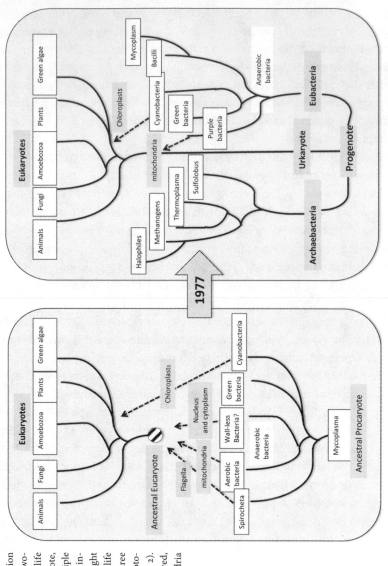

FIGURE 1.6. The Woesian revolution (after Woese, 1981). Left panel: the two-stage rocket scenario. All modern life originated from an ancestral prokaryote, and eukaryotes originated from multiple endosymbiosis events (*dotted lines*) involving various prokaryotic species. Right panel: Woese's scenario. All modern life originated from a progenote with three major lines of descents, including a proto-eukaryotic one (urkaryote; see chapter 2). Only two endosymbiosis events occurred, leading to the emergence of mitochondria and chloroplasts.

Eukaryotes were no longer the descendants of modern prokaryotes, but one of these lineages. This was tentatively a scientific revolution, which would lead to impassioned debates among specialists in evolution.

On a larger scale, as more and more laboratories gradually adopted techniques for studying evolution on a molecular scale, similar to those put in place by Woese and his team to retrace the evolution of organisms, scientists started questioning many traditional classifications. Thus "evolutionists" today no longer use terms such as "plants," "algae," "or protists," because they have realized that the organisms grouped under those names did not correspond to true families from which all individuals descended from the same common ancestor.

Thanks to Woese's work, the study of microorganisms underwent a true revolution. Ribosomal RNA became the Rosetta stone that enabled biologists to identify large microbial groups and to question many established ideas. Microbiologists had become used to classifying bacteria between Gram-positive (Gram$^+$) and Gram-negative (Gram$^-$) according to artificial coloring—blue or pink—carried out in the laboratory. The Gram coloring is fixed in a stable way on bacteria whose wall is rich in peptidoglycan, a giant macromolecule that forms a rigid envelope around almost all bacterial cells and serves as protective armor. This Gram$^+$/Gram$^-$ classification would prove to be misleading on an evolutionary level; indeed, the Gram$^-$ eubacteria could belong to groups of eubacteria that are very different from each other, according to their ribosomal RNA. Similarly, certain methanogenic archaea are Gram$^+$ because, as we have seen, they possess a thick wall containing a molecule that resembles a peptidoglycan. The Gram$^+$/Gram$^-$ distinction thus cannot serve as an evolutionary marker.

Another accepted idea is this: in Woese's time it was believed that the smallest known bacteria, mycoplasmas, were also the most primitive, and thus the most ancient. These bacteria without walls, which had to be Gram$^-$, live inside the human cells they infect. Analysis of their ribosomal RNA has shown, however, that mycoplasmas were not primitive bacteria, but had evolved from more complex Gram$^-$ bacteria that had lost their walls to facilitate their exchanges with parasited human cells.[9] Thus evolution did not always go from simpler to more complex; it sometimes went in the other direction, from more complex to simpler.

Besides correcting some major misconceptions in the way microbiologists used to classify microorganisms, as we will see in chapter 4, the use of data obtained with ribosomal RNA would enable a new discipline, microbial molecular ecology, to develop. It became possible to identify in the field the presence of microbes, thanks to the signature of their ribosomal RNA, which in the past twenty years has enabled scientists to discover the extraordinary abundance and diversity of microbes in nature, which in turn has led to the current interest in our microbiome.

The method Woese used, based on ribosomal RNA, still had significant limits; it did not deal with viruses that, as we've seen, do not possess their own ribosomes. But this was of little concern at the time, because for most biologists viruses were simply fragments of prokaryotic or eukaryotic genomes that had escaped from their host cell a long time ago by acquiring a capsid to form a virion and reinfect another cell. Viruses were in some way considered to be the subproducts of evolution, not of great interest to evolutionists, who were seeking to retrace the grandiose history of life on our planet. We will see below that this simplistic view of viruses is questioned today. But let's look back to the time when Woese had just discovered the original nature of methanogens. What do they have to do with our microbes from hell?

ARCHAEA: THE MOST INFERNAL MICROBES FROM HELL

The Good News Spreads

To elevate microbes that produce methane to the rank of primordial kingdom, on the same level as bacteria and eukaryotes, posed a problem: How could one confer this prestigious status on a single group of organisms, knowing the extraordinary diversity of eukaryotes and the no less great diversity (even if it was less evident at the time) of classic bacteria, the eubacteria? The answer was simple: there must also exist a great variety of archaebacteria. One had only to find them. So Woese and his team set off without delay to search for new archaebacteria, basing their work on what was known about methanogens at the time. Could they identify a characteristic in methanogens that might guide their search for related organisms? They quickly found Ariadne's clew; the methanogens showed an unexpected natural resistance to antibiotics known to inhibit the growth of all pathogenic eubacteria. For example, they were resistant to penicillin and related antibiotics. Penicillin is known to prevent the biosynthesis of peptidoglycan, which leads to the death of eubacteria. In fact, in contrast to most bacteria known at that time (with some exceptions; see below), methanogens do not possess peptidoglycan, which explains why they are resistant to penicillin. Methanogens were also known to be resistant to antibiotics blocking the functioning of eubacterial ribosomes. In the end, it wasn't surprising that antibiotics didn't act on methanogens, since the latter have archaebacterial, not bacterial, ribosomes.

By delving into the literature, Woese identified a small number of other "bacteria" that were also known to resist antibiotics naturally. In 1978 Woese and his colleagues were initially able to show that the 16S ribosomal RNA of a bacterium called *Halobacterium halobium*, known for a long time for its ability to live in hypersalty environments (e.g., the Dead Sea, salt marshes), greatly resembled that of methanogens. Halophilic bacteria were thus also archaebacteria.[10] Unlike methanogens, *H. halobium* and related species (at the time called halobacteria, "bacteria that like salt") need oxygen to live (they are called aerobic). The grouping of these microorganisms with methanogens in the new domain of archaebacteria validated the idea that a third group of living beings on Earth were also characterized by a large variety of organisms and modes of life.

A year later, in 1979, Woese's team would discover a third type of archaebacteria—the thermoacidophiles—by proving that two microbes discovered a few years earlier by Thomas Brock in Yellowstone, a moderate thermophile, *Thermoplasma acidophilum*, and the extreme thermophile that I already mentioned, *Sulfolobus acidocaldarius*, were related through their 16S ribosomal RNA to the methanogen archaea and to *Halobacterium halobium*. These two microbes from hell were also resistant to antibacterial antibiotics. *T. acidophilum* had been isolated from slag heaps of a coal mine in which there had been combustion reactions releasing heat. It multiplied in the lab in a very acidic environment (pH =1) and at 60°C.[11] *S. acidocaldarius*, which came from Yellowstone, was a bit less acidophilic than *Thermoplasma* (its preferred pH was around 3), but it broke all thermophile records at the time with a maximum growth temperature of 85°C. These microbes were aerobic, just like *H. halobium*, and they acidified their environment by producing sulfuric acid thanks to the oxidation of elementary sulfur.[12]

The elevation of *T. acidophilum* and *S. acidocaldarius* to the status of archaebacteria would give these two microbes from hell a very specific status in the eyes of microbiologists, many of whom were unaware of their existence until then. In particular, the fact that *S. acidocaldarius*, the "hottest microbe in the world," belonged to a completely new "primordial domain" of the living world—made them want to know more. Could they find other microbes that were even more thermophilic than *S. acidocaldarius*? Would these also prove to be archaebacteria? If yes, why? And would that teach us something about the birth of life on primitive Earth?

It took some time before Woese's discoveries would be recognized for what they really were and begin to truly influence the scientific community. At the time of the events we've been discussing, the end of the 1970s, very few molecular biologists had heard of Woese and his archaebacteria. Science today is very compartmentalized; a specialist in genetics or in molecular biology might know nothing of the advances in microbiology, except for the few tricks he or she needs to know to culture in the lab the model bacterium used for experiments.

But sometimes it just takes a personality to change the face of the world, and the world of the sciences is no exception. In this case, the personality was a famous German scientist, Otto Kandler, a specialist in the chemistry of bacterial cell walls. He had just discovered that the wall of certain methanogens—which, like everyone else at that time, he still took for bacteria—did not contain peptidoglycan, and this greatly surprised him. A group of methanogens, the Methanobacteriales, did possess a sort of peptidoglycan, but it was chemically different from that of bacteria. Even more interesting, the other methanogens didn't have a wall at all, even though their shapes resembled those of classic bacteria.[13]

At the time, the culture of methanogens, very strict anaerobes, was not an easy thing. To obtain samples and study their envelopes, Kandler collaborated with American microbiologist Ralph Wolfe, who was the first to succeed in obtaining methanogens in pure cultures. By chance, Wolfe's lab was located in the same institute as Woese's. This enabled Woese to extract and analyze 16S ribosomal RNA from four methanogens in 1977. While visiting Wolfe's lab to discuss their collaboration, Kandler learned that Woese had just shown that methanogenic "bacteria" were not bacteria but archaebacterial. For Kandler, this was a revelation. He finally had an answer to his question: Why do the majority of methanogens not possess walls, and why do those who have them not contain the classic peptidoglycan of bacteria, but a modified form of it? Thus it was that Kandler became Woese's first disciple. And it was Kandler who brought the good news to Europe: there exist not two, but three domains of living beings on Earth. These three great scientists—Carl Woese, Ralph Wolfe, and Otto Kandler—became good friends, continuously supporting each other and hiking together in the mountains when they got together at scientific meetings (Figure 1.7).

85° . . . 93° . . . 97° . . . 103° . . . 110°C: Heat Waves in Hell

Back in Germany, Kandler told his biologist colleagues about Woese's discovery, which explains why from that time that country became the home of archaebacteria. Kandler's news in particular drew the attention of another biochemist, Wolfram Zillig, who was immediately smitten with archaebacteria. Zillig was already very well known on the international scene for his work on essential enzymes, RNA polymerases (those that transcribe the DNA message into messenger RNA; see figure 1.3). Quite naturally, he began studying

FIGURE 1.7. Three pioneers of archaeal research. From left to right, Ralph Wolfe, Carl Woese, and Otto Kandler (courtesy of Maya Kandler) during a hiking trip in Bavaria following the first meeting on archaebacteria held in Munich in 1981.

archaebacteria at the end of the 1970s by purifying the RNA polymerases of two of them, *Halobacterium halobium* and *Sulfolobus acidocaldarius*, to verify that their polymerase RNA were different from those of eubacteria.[14] The results were spectacular: not only were they very different from eubacteria (for example, they were resistant to rifampicin, an antibiotic used to treat tuberculosis that inhibits bacterial RNA polymerases), but they were more complex. Surprisingly, the RNA polymerases of archaebacteria bore a closer resemblance to those of eukaryotes (like humans) than of bacteria. The RNA polymerases of bacteria are formed by the association of four proteins (which we call subunits), whereas those of archaebacteria and eukaryotes are formed from a dozen subunits. Up until then, it was thought that the RNA polymerases of eukaryotes were more complex because eukaryotes are "higher" organisms compared to prokaryotes. And then it turned out that the archaebacteria—prokaryotes, since they don't have a true nucleus—had RNA polymerases as complex as ours. This was completely unexpected.

And it was not just a matter of resemblance. While collaborating with the laboratory of André Santenac in Saclay near Paris, Zillig would show in 1983 that antibodies directed against the RNA polymerase of yeast (a eukaryote) recognized the RNA polymerases of archaebacteria, but not those of eubacteria.[15] This was the first in a series of discoveries that would reveal a strong kinship link between the molecular mechanisms of archaebacteria and our own (those of eukaryotes). For example, not long afterwards, in 1984, I discovered that the replication of DNA in archaebacteria was inhibited by a drug, aphidicolin, known to inhibit the polymerase DNA of eukaryotic cells.[16]

These discoveries explain why, in 1990, Woese decided to rename archaebacteria just "archaea," so as to no longer confuse them with bacteria, and to rename eubacteria "bacteria."[17] Indeed, if archaea superficially resemble bacteria, when we scratch the surface and are interested in the great molecular mechanisms characteristic of the living, we notice that archaea resemble eukaryotes more than they do bacteria. In addition, the construction of a universal tree of life, which we will discuss in chapter 4, suggests that this resemblance indeed reflects an evolutionary kinship: archaea would be our sisters, and bacteria our cousins. (From now on, I will use the term "archaea" in place of the now outdated "archaebacteria," and I will no longer speak of "eubacteria," but of "bacteria.") In the same 1990 paper, Woese and his colleagues introduced the term "domain" to replace "primary kingdoms." Again, this name is now commonly used by most biologists, and I will use it from now on in this book.[18]

We will see in the final chapter that Zillig's first observations of the close kinship between archaea and eukaryotes today lead to fundamental—and controversial—questions concerning our origin: Do we share a common ancestor with archaea, or are archaea our direct ancestors? But let's go back to our friend Zillig. From the beginning of his career, he had remained locked up in his laboratory purifying polymerase RNA. That must have weighed on him, because he was a great athlete and traveler; he adored the mountains, glaciers, and nature in general. He was offered the opportunity to leave the confinement of the laboratory to explore hot springs in the field through a surprising—to him—study published by his colleagues in Naples. In 1976 they had purified the RNA polymerase from a strain that seemed related to *Sulfolobus* and that they had called *Caldariella*. That new microbe from hell had been isolated in hot springs near Solfatara, a small town not far from Naples. Unlike Zillig, they had found that this polymerase RNA was typically bacterial, with five subunits instead of ten for the RNA polymerase of *Sulfolobus acidocaldarius* that Zillig isolated in 1979. Zillig wrote to the Naples scientists to request the strain of *Caldariella* so he could study its RNA polymerase himself, but the package containing the strain was lost between Naples and Munich. To ease his mind, Zillig decided to go to Solfatara himself to isolate his own *Caldariella*.

Zillig—who was already over fifty at the time—was accompanied by Karl Stetter, a promising young microbiologist who had done his doctoral thesis under Kandler (the circle is complete). The association between these two forces of nature would prove to be amazingly effective. Although Zillig was a biochemist of international renown, he had no experience in microbiology in the field: he had never isolated the slightest microbe. Stetter was already an expert in that area. They had no difficulty finding and isolating *Caldariella*, which proved to be a close relative of Brock's *Sulfolobus acidocaldarius*. They decided to call it *Sulfolobus solfataricus*. When they were back in Munich, they purified the RNA polymerase of *S. solfataricus*, which turned out to be completely similar to that from *S. acidocaldarius* (a complex enzyme of eukaryote type with ten subunits).[19]

This detour by our two comrades through the Solfatara Caldera was to have considerable historical importance. Imagine the ambience of that place. Half the caldera is wooded—you can even camp there; the other half is a barren field of grays and yellows out of which rises silver smoke in the light of the rising sun, and where one strolls through bogs of boiling gray mud. On the sides of the caldera, sulfurous smoke escapes from the many fissures. What causes such a spectacle? Although the volcano no longer produces lava, its activity is seen in the emission of fumeroles whose temperature varies from 95 to 150°C and are rich in water vapor, carbon gas, and hydrogen sulfur (thus the odor of rotten eggs that awakens those who camp there). When the gases mix with the volcanic ash that came from past eruptions, this causes the muddy bogs. The ambience of Solfatara and the ease with which they had isolated their own *Sulfolobus* would lead Zillig and Stetter to make a big decision. They were hooked on that infernal place. They would set off to explore the world in search of new hot springs and new infernal creatures. Yes, but would they really discover new organisms or simply repeat here and there the work of Brock? Would they simply collect new *Sulfolobus*?

Brock had focused on a hot springs in Yellowstone and he had found *Sulfolobus*, but he had made no particular effort to do that: he had isolated the first microbe that had agreed to grow in his "boiling culture"—a sample of water from that spring, placed in a nutritive liquid medium and brought, in the lab, to the very high temperature with the acid pH of its native source. In particular, Brock had drawn his samples in the presence of oxygen (the easiest method). He had thus isolated aerobic organisms (*Thermoplasma* and *Sulfolobus*). Zillig and Stetter decided to change things up by searching for anaerobic thermophiles. This approach was more difficult, because they had to draw samples from hot springs while being very careful to place them in bottles whose atmosphere contained no oxygen. But their efforts paid off: if archaea other than *Sulfolobus* lived in hot springs (which remained to be verified), why wouldn't some of them be anaerobic, just like methanogens? Let's remember that it was postulated at the time that archaea were very ancient microbes. If thermophile archaea had existed on primitive Earth—whose atmosphere, as we know, was deprived of oxygen—they had to have been anaerobic. Were their descendants still present in the hot springs that are scattered over our planet?

Where to start? Stetter and Zillig decided to begin their quest in Iceland, the island where fire and ice cohabitate, where smoke escapes from banks of snow and one can have midnight baths in pools carved from lava (figure 1.8).

Iceland is in fact marked by extremely active volcanoes, which is seen in particular by the presence of many geysers and hot springs. For a lover of mountains like Zillig, these infernal landscapes were paradise.

Stetter and Zillig's first Icelandic expedition took place in 1980, only three years after Woese's "discovery" of archaea. The cultures achieved in the absence of oxygen enabled them to isolate four new types of infernal microbes very different from *Sulfolobus*. Their first catch was a new methanogen archaea that they named *Methanothermus fervidus*. We already knew of moderate methanogen thermophiles, such as *Methanobacterium thermoautotrophicum*, whose optimum temperature was 65°C, and that still grew at 75°C.[20] But *M. fervidus* had an optimum temperature of 83°C and still grew at 97°C, close to the boiling point of water (whence its Latin name, *fervidus*, which can

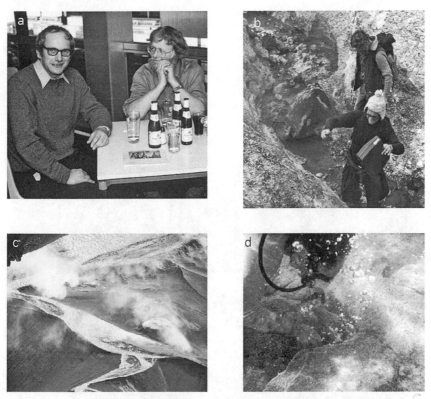

FIGURE 1.8. German biologists Zillig and Stetter hunting for hyperthermophiles. *A*, Stetter (*left*) and Zillig (*right*) at the Reykjavik airport in Iceland during their historic trip in 1980. *B*, Zillig (*top*) and Stetter sampling in Iceland. *C*, Icelandic smoking mountains. *D*, Stetter sampling during a dive near Vulcano island shore. Courtesy of Karl Stetter (*A, B, D*); personal collection (*C*).

be translated as "boiling"). The record for *Sulfolobus solfataricus*, 85°C for its maximum growth temperature, had been broken.

The second catch would prove to be even more interesting. This involved strange organisms in the shape of a stick that multiplied curiously, giving birth to branched shapes in which spheres appeared to be attached to the sticks (figure 1.9a).

They were named *Thermoproteus* (genus name) in reference to the Greek god Proteus, who was capable of assuming multiple shapes. You can see a resting one in figure 1.9b, looking like a classical *Bacillus*. The first species described received the nickname of *tenax*, "tenacious" in Latin, to honor its resistance to high temperatures. *Thermoproteus* was not a methanogen or a thermoacidophile, because it lived in hot springs with a neutral pH. Its optimum growth temperature was 88°C, higher than that of *Methanothermus*, like which it still grew at 97°C.[21]

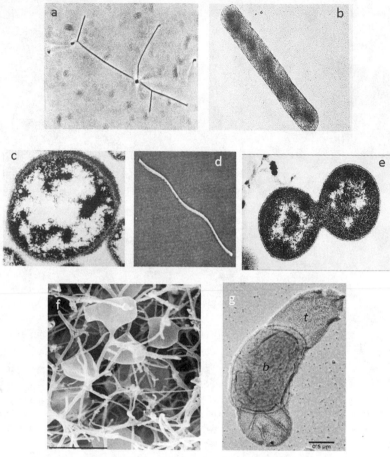

FIGURE 1.9. The zoo of hyperthermophiles. A, optical micrograph and B, TEM of *Thermoproteus tenax*. C, TEM of *Desulfurococcus mobilis*. D, SEM of *Thermofilum pendens*. E, TEM of *Thermococcus celer*. F, SEM of *Pyrodictium occultum* cells connected by canulae. G, TEM of the hyperthermophilic bacterium *Thermotoga maritime* with the cell body (c) and the toga (t). Courtesy of Wolfram Zillig (*A–E*), Karl Stetter (*F*), and Évelyne Marguet, Université Paris-Saclay (*G*).

The third type of anaerobic thermophile organism discovered in Iceland corresponded to spherical cells that, at first glimpse, could have been confused with *Sulfolobus* (figure 1.9c). But instead of transforming sulfur into sulfuric acid, like *Sulfolobus* did, these organisms transformed it into hydrogen sulfur (H_2S), which, for chemists, corresponded to a reduction. They were thus named *Desulfurococcus*, which means "the shell that reduces sulfur" (the prefix *de* indicates a negative action, such as in the verb "detach"; the chemical phenomenon of reduction corresponds to the "detaching" of electrons from an atom).[22]

Finally, they isolated microbes that resembled neither a sphere nor a stick, but long and thin filaments, which they named *Thermofilum*, "the hot filament" (figure 1.9d).[23] It also divided without any problem at 90°C. *Thermoproteus*, *Desulfurococcus*, or *Thermofilum* could have been bacteria or archaea, or could even have belonged to new groups that were unknown up to then. Back in their lab in Munich, Stetter and Zillig purified the polymerase RNA of these three organisms to see what they were dealing with. To their great delight, in each instance these RNA polymerases resembled that of *Sulfolobus*. In all three cases, they were indeed dealing with new archaea.[24] They then sent samples of these microbes from hell to Woese so he could analyze their genetic imprints, which confirmed that they belonged to this new group. The domain of archaea was thus enriched with new phenotypes, alongside (anaerobic) methanogens, (aerobic) halophiles, and (aerobic) thermoacidophiles; they now also had to deal with neutrophilic anaerobic "extreme" thermophiles (living with a neutral pH).

Their first try was a huge success. The following year, in 1981, our two microbiologists again went to Italy, to Naples, but this time took the night boat. They went to the Aeolian Islands, and, more precisely, to Vulcano, an island made up of two volcanoes: a small one, Vulcanello, now extinguished, and a larger one, Vulcano, an active volcano whose last eruption was in the nineteenth century. A fissure descends from Vulcano's crater and goes to the sea, a few meters from the beach where vacationers enjoy themselves. And so there are a number of hot springs on the beach, testimony to the volcanic activity of the place. Looking out to sea, one occasionally sees bubbles on the surface, which appears to be agitated by eddies. These are created by gas rising from the magma deep below (carbon gas, in particular), out of underwater hot springs. On the beach, in those holes of boiling water, Stetter and Zillig discovered another anaerobic microbe, *Thermococcus celer* (literally, "the thermal shell that moves quickly," from *celer*, "speed" in Greek), capable of growing in temperatures up to 93°C (figure 1.9e).[25] An analysis of their RNA polymerase and their ribosomal RNA would show that *Thermococcus* was also an archaea, but was very removed from *Desulfurococcus* and *Thermoproteus*. In fact, it seemed closer to the methanogen and halophilic archaea; this was the first indication of the existence of two great branches (or phyla) among archaea.[26]

This was only the beginning. The island of Vulcano had fascinated our two scientists. The following year, in 1982, they returned to the site, but to different locations. In the meantime, Stetter had been recruited to teach at the University of Regensburg. He quickly had a large lab wonderfully equipped to study the organisms he had just discovered. He now had to prove himself and emerge from Zillig's shadow. First, Stetter discovered a microbe that would become a reference in the world of hyperthermophiles, *Pyrococcus furiosus* ("the furious burning shell"). Indeed, it moved around furiously in every direction under the microscope, no doubt to warm up, using its many flagella.

This archaea (because of course it was still an archaea) is very close to *Thermococcus*, but it exceeded all other thermophiles, with an optimal growth temperature of 95°C. Thirty years after its discovery, *Pyrococcus furiosus* is today manipulated in the lab of Mike Adams in Athens, Georgia; his goal is to produce biofuel from CO_2.[27] Airplanes will perhaps one day fly and cars run thanks to the energy produced by this microbe from hell when fossil fuels will have disappeared.

But let's go back to 1982. That was the year Stetter won the jackpot, catching in his nets another archaea that was even more infernal than *Pyrococcus*, capable of dividing at temperatures above 100°C. During his exploration of Vulcano, Stetter had noticed the existence of underwater hot springs. To collect his samples, he took his family vacation on that island the following summer and brought along his diving equipment. He was thus able to collect samples from hot springs located ten meters below the surface, where the water remained in a liquid state (figure 1.8d). Recall that the temperature at which water boils, which is equal to 100°C at the ambient atmospheric pressure (an atmosphere), increases with the pressure that itself augments with depth (a Bar every ten meters). At a depth of ten meters, the boiling temperature of water is thus 110°C. Back in his new laboratory in Regensburg, Stetter would isolate from his samples an archaea close to *Sulfolobus* and *Thermoproteus*, but whose optimal growth temperature was 105°C, and that still divided at 110°C under pressure. Yet at 90°C, the microbe became immobile; it was frozen. Stetter would name that archaea, which was characterized by its shape and its atypical behavior, *Pyrodictium* ("burning fingers"). In fact, in some photos taken with an electron microscope, it resembled a hand with fingers. Even more surprising, the cells of *Pyrodictium*, which have the appearance of flat disks, are connected to each other by long filaments, the whole almost evoking networks of neurons (figure 1.9f). *Pyrodictium* would hold the record for life in hell until 1993.

The discovery of *Pyrodictium* in 1982, then its published debut in the prestigious journal *Nature*, would establish Stetter's international renown and make him the leader of research in this area.[28] These microbes from hell needed a name that would distinguish them from Brock's earlier children, whose optimum growth temperatures had never gone above 80°C. As in the case of Woese with the term "archaebacteria," Stetter wanted a new name that would perpetuate his discovery. He would invent the name "hyperthermophile" for all microbes whose optimal growth temperature is equal to or above 80°C.[29]

Life at 250°C?

Although Stetter and Zillig's discovery of hyperthermophiles caused a big stir in the small world of microbiologists, the media remained silent on it. In fact, our two scientists were victims of bad luck. Their discovery was eclipsed by a scoop that would prove to be one of the largest scientific blunders of the twentieth century. In 1982, the same year that Stetter discovered *Pyrodictium occultum*, newspapers throughout the world announced the discovery of bacteria capable of dividing at 250°C in dark underwater hydrothermal chimneys (this is the word used in the scientific literature).[30] And so 110°C seemed paltry. It wasn't a tabloid that described these incredible microbes for the first time but the journal *Nature*, the same one that had published the discovery of *Pyrodictium*. And so this report proved at first glance to be serious. The authors of the work in question were two American scientists who had carried out their experiment in the laboratory of Aristide Yayanos, a specialist in microbes that live under high pressure in large underwater trenches.

Deep hydrothermal vents, the famous black smokers, had been discovered a few years earlier (in 1977) by the American geologists John Corliss and Jerry van Andel, thanks to the submersible *Alvin*.[31] These large, tall chimneys are formed by the precipitation of minerals present in the boiling waters that escape from the vent. These minerals, on contact with the cold water of the ocean (2°C), precipitate and form superb black chimneys out of which black plumes of smoke emerge. These "oases of life," to use the preferred expression, immediately fascinated scientists (we will visit these again in the next chapter): whereas the great ocean depths are generally mostly deserted, these deep hydrothermal vents are covered with a multitude of organisms, each more astonishing than the last (giant mussels and shrimp, huge white worms ending with flamboyant red feathering, and many others). They aren't found just anywhere in the depths of the oceans, but near the ridges, those hilly and fractured zones of the ocean floor (usually located at depths of between 2,000 and 3,000 meters) that, when probed, reveal magma at a shallow depth. Seawater penetrates the cracks, is heated by the magma, and then comes back out under pressure through the black smokers.

Let's return to our microbes. At 3,000 meters below the sea, the pressure is that of 300 atmospheres. In these conditions, water remains liquid up to 300°C. Why then reject the idea of "superhyperthermophilic" microbes in liquid water at 250°C? Let's recall that biochemists already had a great deal of difficulty imagining stable proteins at above 50°C. If the discovery of bacteria living at 250°C had been confirmed, we would have had to imagine another form of life on Earth—different molecules. However, in the photo of one of these "superhyperthermophiles" published in *Nature*, one sees a cell that in every way resembles that of a classic bacterium. Were there then new principles of organization of living molecules, new types of chemical links? The existence of bacteria living at 250°C would unlock journalists' imaginations. How far could one go? Why not life in magma? And wouldn't all the stars burning in the sky also be homes to infernal lives, each more amazing than the one before?

The media bubble of bacteria living at 250°C would soon burst. Only two years after the appearance of the paper in question, Jonathan Trent, a brilliant young American scientist, tried to reproduce the culture of these superbacteria from the same initial samples (extracted from liquid taken at the mouth of the black smokers; we'll see how later) in Yayanos's lab. Nothing grew in the medium cultured with this fluid, then incubated at 250°C under a pressure of 300 atmospheres. However, Trent saw the appearance of lumps formed by the accumulation of organic matter present in the medium to be used as food by potential bacteria. These aggregates were highlighted with a dye called DAPI, which is normally used to observe DNA molecules. When you say "DNA," you say "living being"; this is why the authors of the article that appeared in *Nature* believed they saw microbes appear in the medium. What about the photos of bacteria that accompanied the article? There was a logical explanation for that. Trent noticed that one of the products the authors of the contentious article had used to prepare their samples for observation under an electron microscope had been contaminated by bacteria. The "superhyperthermophilic" microbes were only simple colibacilli that had gone astray.

Trent's paper describing the above results was also published in *Nature* in 1984.[32] But it was not picked up by the international media. Why? It's hard to say. Perhaps people simply don't want their dreams to be shattered. In any event, the story of bacteria living at 250°C had two unfortunate consequences. First, the lack of publicity given to the debunking of it meant some people, and not the least intelligent, still think that such bacteria exist. The late American biologist Stephen Jay Gould even cited them in a popular book published in 1997 (excellent, by the way): *Full House: The Spread of Excellence from Plato to Darwin*. Next, the discovery of true hyperthermophile microbes still dividing at 110°C went almost unnoticed by the public at large. But this episode did attract some scientists' attention to the issue of the stability of organic molecules at high temperatures, a very important point to which we will return. In particular, in 1984 the American biochemist Robert White published, again in *Nature*, a paper that resolved the polemics.[33] Placed in conditions that were supposed to be favorable to superhyperthermophiles from the black smokers—300 atmospheres and 250°C—molecules typical of living organisms were rapidly degraded. Specifically, the ATP molecule, which serves as currency for all the exchanges of energy in all cells, disappeared in a few seconds, and proteins in a few minutes. The black smokers (or white, in some cases, depending on the chemical composition of the smoke that escapes from these stacks) are indeed inhabited, but by completely classic archaea. Scientists found at the bottom of oceans—at great depths—little brothers of the archaea that Zillig and Stetter had found in the terrestrial hot springs in Iceland and on Vulcano. They gave them species names that recall their place of birth, such as *Pyrococcus abyssi*, isolated by a French team, which we'll discuss below, or *Pyrodictium abyssi*, isolated by Stetter himself following his participation in an American expedition with the submersible *Alvin*.

THE ZOO OF ARCHAEA FROM HELL EXPANDS

From 1982 until the beginning of the 2000s, Stetter, Zillig, and their colleagues, including many students, pursued their work in the field and continued to enlarge the zoo of anaerobic and hyperthermophilic archaea. These were found not only in terrestrial and submarine hydrothermal vents, but also deep in the terrestrial crust and hot reservoirs.[34] New, evocative names were added to the earlier list: *Staphylothermus*, *Pyrobaculum*, *Ignicoccus*, *Hyperthermus*, and others. Stetter's and Zillig's imaginations were sorely tried. To name their favorite microbes, they had to juggle with Greek or Latin words evoking fire (*pyro*, *igneus*) or heat (*thermos*), and those that characterized their new microbes by their morphology (*baculum* = stick, *filum* = filament, *staphylo* = small chain, or even *coccus* = shell). Thus *Pyrobaculum islandicum* could be translated as "the burning stick of Iceland." In 1986 Robert Huber, one of Stetter's most faithful colleagues, isolated from a deep sea hydrothermal vent an archaeon that is both a hyperthermophile and a methanogen— thus anaerobic—capable of dividing up to 110°C. This fascinating methanogen was named *Methanopyrus kandleri* in honor of Otto Kandler, Stetter's thesis director.[35] The record held by *Pyrodictium occultum* was tied. In 1993, in an underwater chimney off the shores of Iceland, a thesis student from Stetter's lab, Elisabeth Blöchl, discovered *Pyrolobus fumarii* (literally, "the burn-ing lobe of the chimneys"), an archaea whose optimal growth temperature is 106°C and that still divides at 113°C.[36] The record had been broken again!

Is *Pyrolobus fumarii* "the hottest" microbe known on planet Earth? The question was raised following the 2003 publication of a paper in *Science* in which an American laboratory wrote of an archaea whose optimal growth temperature was 106°C, like *Pyrolobus fumarii*, but that could continue to divide up to 121°C.[37] Except Stetter didn't believe the results because he noticed that the method the authors used to measure the growth of the organisms—counting the cells by optical microscopy in a medium rich in iron particles looking like cells—is not reliable and can lead to serious errors in estimation. I know him well, and it was certainly not out of jealously that he refused to give up the 113°C record. I too have many doubts, and most of my col-leagues in the area are also very skeptical because twelve years after publica-tion this result has never been reproduced. This was not the end of the story; in 2008 a Japanese laboratory published in the no less prestigious journal of the American Academy of Sciences (*Proceedings of the National Academy of Sciences*) an article in which it claimed to have grown under pressure a well-known archaea, *Methanopyrus kandleri*, up to 122°C,[38] one degree more than the Americans' record. Up to then, the maximum growth temperature re-corded for *M. kandleri* was 110°C. Is it possible that the increase in pressure is enough to make a leap of 12°C? Here too I have my doubts, because no one has reproduced this experiment to date.

I must note that in all the cases we've been reviewing, the optimal tem-peratures of these hyperthermophiles remain close to 105 to 106°C. These are sure values; the maximum temperatures are much more difficult to determine because microbes from hell, regardless of their taste for high temperatures, do not grow well beyond 108 to 110°C. It is true that some proteins can remain active after being exposed for several minutes in an autoclave at 120 or even 140°C, but we will see in chapter 3 that the same is not true of biological membranes that represent the Achilles' heel of living beings at high tempera-tures. I predict that the record of 106°C for an optimal growth temperature will not be usurped anytime soon.

Up to now, all the microbes from hell that Stetter and Zillig discovered proved to belong to the third domain of the living: archaea—and bacteria? Does *Thermus aquaticus*, which Brock isolated in 1967, still hold the record of maximum growth temperature (80°C) for bacteria? The answer is no. Stetter again went beyond that. In 1984, Huber and Stetter discovered on Vulcano an anaerobic hyperthermophile bacterium, *Thermotoga maritima*, which grows up to 90°C.[39] This microbe is enveloped by a supple wall that detaches from its main body (b in figure 1.9g) at its two extremities, forming structures that look like bubbles (t in figure 1.9g). For Stetter, this called to mind the toga of a Roman senator that does not stay perfectly on his body. In 1989 the same team isolated another anaerobic bacterium that was even "hotter": *Aquifex pyrophilus* (literally, "the microbe that manufactures water and likes heat").[40] Its optimal growth temperature is 85°C, and it can still divide at 95°C.

For twenty-five years, *Aquifex*'s record has held for bacteria, and no new really hyperthermophilic bacteria have been discovered. So hyperthermophilic bacteria exist, but they are fewer and less diverse than hyperthermophilic archaea. Also, they don't come close to reaching their record, since the optimal growth temperature for *Aquifex* (a bacterium) is 20°C lower than that of *Pyrolobus* (an archaeon): 85°C versus 105°C. Why is there this difference between archaea and bacteria? We will discuss this important question in detail in chapter 3.

In the above paragraphs, I've spoken a lot about the work in Stetter's lab in Regensburg. However, Zillig was not sitting on his hands in Munich while Stetter was breaking records in the world of high temperatures. We owe a major discovery to Zillig: that of the first viruses from hell. There's no point trying to hide in a hot springs to avoid our friends the viruses. Even there, they'll find you. Zillig wanted to isolate viruses to use them as study models for the molecular biology of archaea, which the pioneers of molecular biology had done in France and the United States in the 1950s to 1970s. It was in particular thanks to a virus infecting a bacterium (the bacteriophage T4) that it was definitively shown that DNA corresponds to genetic material, or even that the genetic code could be read by three-letter words.[41]

Zillig expected to find phages of archaea producing virions resembling those of bacterial viruses, along with their heads, which contained their DNA, and a tail that enabled them to inject DNA into the cell that they were infecting (figure 1.2d). Indeed, the first archaeal virus discovered in the 1970s, which infected a halophilic archaeon (lover of salt), produced completely classic (head to tail) virions. Imagine Zillig's surprise when, on his return from the first of his Icelandic expeditions, he observed elongated filamentous particles (without a head) attached to cells of *Thermoproteus* (figure 1.10a).

He was able to isolate these filaments and realized that they were indeed viral particles that could infect the strain of *Thermoproteus tenax*. He had isolated the first virus from hell, which he called TTV, for *T. tenax* virus.[42] I can still see today a colored chalk drawing representing the TTV virion on the blackboard of his office in Munich. I almost erased it when I was giving a seminar during one of my visits, thinking it was an old drawing—at which point he leapt up to stop my sacrilegious hand.

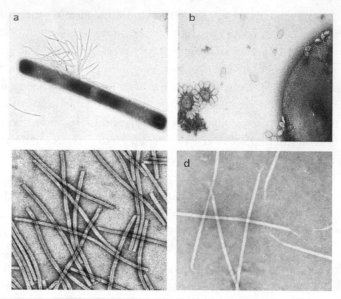

FIGURE 1.10. The first isolated viruses from hell. *A*, TEM of virions of TTV1 attached to *Thermopro-teus tenax*. *B*, TEM of SSV1 virions attached to membrane vesicles and cells of *Sulfolobus shibatae*. TEM of rudivirus (*C*) and lipothrixvirus (*D*) virions. Courtesy of Wolfram Zillig (*A, B*), Emmanuelle Quemin (*C*), and Soizick Lucas (*D*), Institut Pasteur.

The TTV virion greatly resembles that of the tobacco mosaic virus, or TMV, which infects a eukaryote, but the genome of TTV is made up of DNA, whereas the genome of TMV is made of RNA. Thus TTV was very different from viruses known up until then. Unfortunately, TTV has been lost over the years and is no longer available today in any laboratory. However, again in the 1980s, Zillig would isolate from a hot springs in Beppu, Japan, a virus that would become one of the model viruses of archaea. This virus, called SSV1 (*Sulfolobus shibatae* virus 1), infects one of the most studied archaea and produces particles in the form of a lemon or a spindle (called fusellovirus) that are completely different from those of TTV (figure 1.10b).[43] Virions with this shape had never been observed before in viruses infecting bacteria or among eukaryotes.

During new expeditions in Iceland at the end of the 1980s, Zillig would discover new fuselloviruses, but also a virus close to TTV in the shape of a stick (rod-shaped) that infects *Sulfolobus*: SIRV (*Sulfolobus islandicus* rod-shaped virus 1; figure 1.10c).[44] We will look again at this virus, which has become a study model for viruses of archaea in our laboratory at the Institut Pasteur. It is the prototype of a family of viruses that today are called rudi-viruses, or rigid viruses. Zillig compared the virions of rudiviruses to spa-ghetti. During the same expedition, he discovered a new virus producing virions looking like flexible filaments (figure 1.10d), which he compared to noodles and called lipothrixvirus, lipidic filaments, because their virions contain lipids.[45]

During one of his Iceland expeditions, in 1995, Zillig was accompanied by a Georgian scientist, David Prangishvili, who had joined his team in 1992 upon the fall of the Soviet Union. Prangishvili fell in love with the viruses of hyperthermophile archaea and became the uncontested world expert in the field in the 2000s, discovering viruses producing virions in ever more surprising shapes—some resembling bottles or springs, others capable of changing shape after coming out of the cell, still others ending in pincers that they used to grip the flagella (hairs) of the archaea they infected (see figure 2.7). We will talk about Prangishvili and his viruses soon. The discovery of different viruses among archaea, as well as viruses infecting bacteria and those infecting eukaryotes, reinforced Woese's view of a living world divided into three large domains. To each of these domains there indeed corresponded a specific group of viruses. The diversity of viruses from hell, greater than that of bacterial or eukaryotic viruses, was a big surprise for biologists around the world (at least for those who heard about it), again raising new questions about the place of microbes from hell and their viruses in the living world.

What a stunning number of infernal microbes and viruses infecting those microbes our two German scientists and all their colleagues discovered! I wish for a moment to share my admiration for them. The dream of every microbiologist is to describe for the first time a new microbial genus. For microbes, as for other living beings, the genus corresponds to the first term of the Latin name: *Pyrolobus fumarii* is of the genus *Pyrolobus* and the species *fumarii*, just as we are of the genus *Homo* and the species *sapiens*. Genera are then grouped into family and order. Thus the order *Desulfurococcales* contains the genera *Desulfurococcus* and *Pyrodictium*. Viruses are also divided into orders: for example, 95% of viruses infecting bacteria belong to the order of *Caudovirales*, viruses with tails. If you discover a new genus, your academic career is assured (sometimes even a new species is enough). And if you discover two or three new genera, as Brock did, you are a god (for other microbiologists, that is). What can we say in this case about Zillig and Stetter who, in twenty years, were able to describe more than a dozen new genera, several new families, and several new orders of archaea, bacteria, and viruses infecting microbes from hell? Furthermore, these new branches of the tree of life corresponded to completely new organisms on the level of their biological performances (hyperthermophile nature), their metabolism (they can use an extraordinary variety of chemical substances to manufacture their energy), or, for viruses, the structure of their virions. We have just experienced with these two men one of the greatest scientific adventures of the twentieth century. Very few people are aware of this. You are now among the privileged few.

SEARCHING FOR THE ORIGIN OF LIFE

We have already mentioned the challenge that led Stetter and Zillig to set off in search of thermophile microbes. Microbes more thermophilic than those isolated by Brock had to exist; it had to be possible to isolate them if they were carefully cultured in an oxygen-free medium; and those microbes should be archaea. A risky challenge, but they still accepted it. In less than two decades, they revealed a world of microbes that up to then had been completely unknown—that of hyperthermophiles—and with only two exceptions (*Thermotoga* and *Aquifex*), these hyperthermophiles were indeed archaea. The presumed ancient nature of this third domain of the living world was thus reconciled with the absence of oxygen in the atmosphere of primitive Earth. Suddenly, there appeared a correlation between the presumed "ancient" nature of a microbe and its ability to live at extreme temperatures. This connection would strike the imagination of many biologists. What if hyperthermophiles were directly issued from the first cells that had inhabited our planet—cells that would have appeared in the hot springs and the primitive underwater chimneys, whose descendants had continued to live in this Dantean environment? Couldn't we imagine hyperthermophile microbes as witnesses of a past when our entire planet resembled hell?

A THORNY PROBLEM

Once again, the hypothesis implied a very specific interest in microbes from hell. Because the question of the origins of life is one of those that intrigue us the most: Where do we come from? Who are we? Religious myths have given many responses to these questions. To replace those myths, scientists would like to provide a clear and definitive answer. Yes, but then, like all questions that bear on the evolution of the living being, the problem of the origins of life is of a historical nature. And historians depend entirely on archives, testimony, or archaeological finds that enable them to reconstruct the past—all elements that are cruelly lacking in an understanding of the origin of life. We will likely never be able to answer certain key questions concerning the relationships between ancient human empires and kingdoms, because many documents that could have revealed the truth have disappeared, consumed in the dust of the centuries. Similarly, we will perhaps never know how life appeared on Earth. Unless we invent a time machine (unlikely, according to my physicist friends), we are reduced to hypotheses that are more or less credible. And even if one of them enables us to recreate life (or a new life) from nothing (*ex nihilo*) in the lab, we can never be certain that things indeed unfolded that way during the birth of life on Earth.

For a long time, scientists and philosophers thought they could resolve the problem. Life, they thought, had always existed, as had matter. It was continually produced out of inanimate matter: this was the theory of spontaneous generation. Pasteur ruined these beliefs in the nineteenth century by showing that spontaneous generation was just another myth: every cell comes out of another cell. Today, in spite of the difficulties, a small community of scientists is attempting to attack the great problem of the origin of life. A Russian scientist of the past century, Alexander Oparin (1894–1980), is considered to be their spiritual father.[46] It was he who opened the door in 1924, stating that the first forms of life must have developed in an environment that was very different from ours, in particular in the absence of oxygen. He imagined that, in those conditions, organic matter could be formed through the action of solar energy and electrical energy (dissipated during thunderstorms) on the molecules in the primitive atmosphere, which was believed at the time to be essentially made up of methane. The bricks of life thus formed would have accumulated in the oceans of the young Earth to form the famous "primordial soup," where they would progressively assemble and make the first cells "appear."

In 1953 a young American scientist, Stanley Miller, succeeded in producing certain amino acids in the lab by subjecting to electrical charges a glass beaker filled with a gassy mixture whose composition (methane, hydrogen) would mimic that of the primitive atmosphere.[47] But later, things became a bit complicated. The composition of the primitive atmosphere is today a controversial subject. Most scientists think it contained very little methane, but was essentially made up of carbon gas, dinitrogen, and water vapor. In that case, Miller's experiment didn't work. Other scientists have criticized the notion of "primordial soup": it was too diluted, they said, for the ingredients of life to be able to come together in the ocean and turn into macromolecules, then into cells.

FROM MICROBES FROM HELL TO THE ORIGINS OF LIFE?

Let's return to our microbes from hell. Oparin's hypothesis and Miller's experiment are not unlike a recipe for cooking. With most recipes, you have to use heat. Heat transforms food: the beaten egg becomes an omelet; fruit becomes jam. Links between atoms are broken, while others are created; something new appears. Why wouldn't life have appeared in the same way, through heat? In his time, Darwin spoke of a small, warm pond in which primitive organic molecules would have steeped for a long time in the sun before forming the first living cell. With the discovery of microbes from hell, some have said, "What if the pond in question had been boiling?" They thought—and we still think so today, from hard evidence—that Earth was formed some 4.57 billion years ago through the accretion of millions of grains of interstellar matter, types of meteorites that would have assembled following violent collisions, at the origin of increasingly large bodies. Each shock liberated a large amount of energy in the form of heat. The primitive Earth was probably unable to dissipate the colossal heat produced by all those impacts, so that it probably remained in a state of fusion, constituting a sort of ball of magma that later cooled. The primitive atmosphere of Earth was then formed from that cooling and from the degassing that followed. The carbon gas and water vapor that it contained were at the origin of a very intense greenhouse effect, but worse than the one we worry about today, with the emission of carbon gas due to human activity. With Earth being in a sense a prisoner of this infernal atmosphere, the rays of the sun would thus have increased Earth's surface temperature to 200°C, according to some calculations. A burning atmosphere, oceans of fire . . . only a few little demons and a devil were missing. Of course, the composition of the atmosphere later evolved, and at a date corresponding to that of the first fossil forms of life detected in the archives of Earth (three billion years ago), the greenhouse effect and the temperature must have been much lower (we'll come back to this). But with the discovery of microbes from hell, it was tempting to assume that the first forms of life were more ancient and to imagine an infernal primitive Earth populated with hyperthermophiles. Amazingly, Bradley Moore Davis, one of the first two biologists who looked at the Yellowstone hot spring, was already flirting with this idea, writing in 1897, "Perhaps, but this does not necessarily follow, these organisms resemble more closely the primitive first forms of life than any other living types."

The discovery of black smokers at the bottom of the oceans would add some spice to this history. Why? The meteorite shower that I described above only gradually decreased. The very young Earth was thus intensely bombarded. Astrophysicists have calculated that the falling of a celestial body 400 kilometers in diameter would release energy capable of vaporizing almost all the water contained in the current terrestrial oceans, which is largely sufficient to wipe out all forms of life. For them, primitive Earth was thus regularly sterilized by this bombardment. If life appeared on the surface between two falls of giant meteorites, everything had to start all over again after each fall.

When things began to calm down a bit (around four billion years ago), we can imagine that the less frequent falling of giant meteorites was still sterilizing the planet, but only on the surface. The depths of the oceans—which had in the meantime been formed—were thus protected. Unlike the surface hot springs, exposed to the devastating effects of rocks falling from the sky, the deep springs—the famous smokers—were then perhaps true little paradises. They were protected not only from the activity of the meteorites but also from that of the sun's ultraviolet rays. At the time, those rays would have directly struck Earth's surface, since the ozone layer didn't exist: ozone is formed from atmospheric dioxygen, a gas that didn't appear in the atmosphere until much later, around 2.4 billion years ago. And ultraviolet rays are particularly destructive to life.

If it were assumed that life appeared in underwater hydrothermal springs, this explained how the "young shoots" were protected from both the bombardment of meteorites and the intense ultraviolet rays. And in such a scenario, the energy that served to unleash and maintain the interactions of the living no longer came from the sky, but from the earth itself. It was the heat coming from the depths of the earth that, while heating, decomposing, and recomposing molecules, would have enabled the formation of organic matter. Oases of life became the *cradles* of life. In addition, the smokers were capable of providing not only protection and heat but also all sorts of minerals and gases essential to life.

John Corliss, the American scientist who discovered the deep hydrothermal springs in 1977, was the first to formulate the hypothesis according to which life appeared at high temperatures in the smokers in the great ocean depths.[48] As we have seen, those smokers are indeed one of the favorite residences of hyperthermophile archaea. They live in the walls of the hydrothermal chimneys—within which there is a temperature gradient that goes from 300°C in the interior (in contact with the fluid that goes through the stack) to 2°C at the exterior (the temperature of the ocean)—or, more precisely, in the interstices of the porous rock that constitutes the pipe of the chimney. There remained only a step to be taken to consider the presence of archaea—those *assumed* "old bacteria" in environments uniting so many conditions propitious to the appearance of life—to be a decisive argument in favor of the hypothesis of life beginning in high temperatures. And so we can understand why the hypothesis of a direct link between microbes from hell and the origin of life has become so popular among scientists and in the media.

We can go even further if we imagine, along with some scientists, that Earth's crust itself is an inexhaustible reservoir of hyperthermophiles.[49] Indeed, the farther one digs into the ground, the more the temperature rises.

We know, moreover, that microbes can live (or survive) in liquid inclusions present in rocks up to a very great depth. Hyperthermophiles are thus present in this particularly infernal environment. For instance, *Pyrococcus abyssi*, which was isolated for the first time from a deep hydrothermal hot springs off the coast of the Fiji Islands, was later found at the bottom of a South African gold mine. In that case, mightn't life have appeared in Earth's crust itself? That's something that would be of great interest to exobiologists. In such a scenario, one need no longer wonder about the distance separating a planet from its star and about the composition of its atmosphere when looking for other possible cradles of life in the universe. Any rocky planet can suffice: one need only descend under the surface to the right depth to obtain a correct temperature and to image ad hoc chemicals in this deep biosphere to feed possible primitive microbes with gas and adequate mineral salts.

These possibilities enflamed and still enflame a few scientists and some journalists who see life everywhere in the universe. A few years ago we read on the front page of a major French newspaper an article predicting the discovery of "archaebacteria" in hot springs that the journalist imagined were present in the depths of the ocean under the surface of Europa, one of the satellites of Jupiter. This was going a bit too far. Above all, it neglected the fact that archaea are a particular form of terrestrial life. If our descendants one day discover living beings under the crust of the ice of Europa, they will be very different from terrestrial organisms because their history will have been completely different. Among other things, they will need new names.

In 1996, the international meeting on thermophiles held in Athens, Georgia, was followed by a colloquium entitled "Thermophiles: The key for understanding the origins of life?" The question mark was purely a formality: in the work that was published two years later from the acts of this colloquium, all the scientists except one were in favor of the hypothesis of a hot origin of life.[50] I was the holdout. The title of my chapter was "Were Our Ancestors Actually Hyperthermophiles? Viewpoint of a Devil's Advocate." I was indeed the devil's advocate, raising problems posed by that hypothesis. I will explain them in detail in chapter 4. Since then, the situation has evolved quite a bit, in particular thanks to the work of a team of French scientists in Lyon who have undertaken to determine at what temperature our ancestors lived. Their work won a prize in 2012 from the journal *La Recherche*. They proposed an unexpected scenario: The last ancestor common to all current organisms would not have been a microbe from hell, but a "mesophilic" organism like you and me, or a moderate thermophile. By contrast, the last ancestor common to all bacteria, as well as the last ancestor common to all archaea, would both have been microbes from hell. How can this be explained? If life appeared at high temperatures—but this is only a hypothesis—why wasn't LUCA a microbe from hell? And if LUCA lived in a temperate environment, why did the ancestors of the archaea and the bacteria live in hot environments? We will see that there are different interpretations. In all cases, determining the place of microbes from hell in the living world appears crucial to whoever wishes to reconstruct the history of life on our planet.

WHEN MICROBES FROM HELL ISSUE CHALLENGES

In quickly reviewing a few pages of the recent history of microbiology, we have witnessed a complete reversal of the situation. For decades, the observations of the pioneers from Yellowstone at the beginning of the twentieth century remained unknown because biologists were convinced that it was impossible for living beings to function at over 70°C. No one had predicted the existence of microbes from hell. Brock's early work in the 1960s itself remained virtually unknown to the great majority of biologists. A few years later, those microbes whose existence hadn't even been imagined took center stage and found themselves at the heart of one of the most heated debates in biology: the question of our origins.

These microbes from hell will accompany us throughout this book. Alongside them we will attempt to respond to these fascinating questions:

- The question of life in hell: How can life exist and thrive in temperatures above the boiling point of water? How do microbes from hell function to survive at temperatures that "coagulate protoplasm," to repeat Setchell's 1903 question?
- The question of the origin of life: Where do microbes from hell come from? What do they teach us about the history of living beings? Why are the most thermophile microbes all archaea? Why aren't there hyperthermophile eukaryotes? This last question is particularly interesting. At present we don't know of any eukaryotic organisms, microbes or others, capable of living permanently at temperatures above 60°C, which is far from being a truly infernal temperature. This record temperature has been held since the 1960s by fungi that Brock discovered in Yellowstone. There are perhaps hyperthermophile eukaryotes, and Stetter himself actively looked for them, but he didn't find any. Why did prokaryotic cells, archaea and bacteria, which look "primitive," succeed where eukaryotic cells, which are more complex, failed?

An attempt to answer all these questions introduces many challenges that have caused two generations of scientists to hold their breath, and they will not be answered anytime soon (there will still be work for those of our younger readers who would like to join in the adventure). But before trying to answer what I can using the tools of modern biology, I will say a few words about the reasons (and about the vagaries of history) why I am so interested in microbes from hell. We will then set off to meet these microbes in the field, where they live; we will see how it is possible to bring them home to the lab. For this, we will accompany a scientist who explored the depths of the Pacific Ocean on board a small yellow submersible, the *Nautile*, 2,300 meters beneath the ocean. This expedition enabled us to undertake new research that was to change my life and the lives of several of my colleagues. This will allow me to introduce our research teams and tell a story that I hope is far from over.

Hunting Hyperthermophiles and Their Viruses:
From the Great Depths to the Laboratory

I have never been attracted to hell, so I have no reason to be interested in its microbes. When I was nineteen, I discovered the amazing hot springs in Iceland during a hitchhiking and Land Rover trip with a childhood friend, but I would never have imagined that microbes could live in those holes of boiling and muddy water that are scattered across the country. I was happy at sunset to admire the steam that escaped from them, forming lovely multi-colored tableaux. It was the discovery of archaebacteria twelve years later that would lead me to hyperthermophiles. At the time, I was preparing a thesis on the mechanisms that enabled bacteria to replicate their DNA so that each daughter cell receives a copy of the chromosomes from the mother cell. For a model I used the famous bacterium *Escherichia coli*. I got off to a rather bad start because I was working alone on this subject in a small research group in competition with strong American teams, one more excellent than the other. Also, I didn't get along with my supervisor. So I spent more time teaching, being involved in politics (this was the 1970s), and taking care of my young son than conducting experiments in the lab. Luckily, I had a permanent job as an assistant at the university, a uniquely French situation at the time, which kept me from getting fired due to a lack of results and allowed me to wait for better days ahead.

Enter Carl Woese, the Guy Who Changed My Life

I remember well the key moment in the early 1980s that changed my life. I was on vacation with a girlfriend who had a delightful house in the middle of a forest in Landes, a large green expanse that covers a good part of south-western France and culminates on the vast beaches of the Atlantic. I was lying in a hammock as Annie prepared her specialty, baked stuffed oysters. I had brought something to read, and so I opened the most recent issue of *Pour la Science*, the French version of *Scientific American*. I started to read an article with a strange title, "Archaebacteria," by a certain Carl Woese, whom I had never heard of.[1] This was of course a revelation: so there existed prokaryotes, which were not bacteria. Woese explained that a comparison of the molecular mechanisms of bacteria, archaebacteria, and eukaryotes should enable a reconstruction of the ancestor common to all contemporary living beings, which he called the progenote because he envisioned it as a very primitive organism.

Last but not least, Woese criticized the prokaryote/eukaryote paradigm, in particular the idea held at the time (which some evolutionists still hold today, as we will see in chapter 5): that eukaryotes derived from prokaryotes, which at the time were associated with bacteria. As we saw in chapter 1, he illustrated this traditional view of evolution with a diagram that made me think of a two-stage rocket (figure 1.6). In the first stage—that of prokaryotes—an ancestral bacterium gave birth to many lineages that diversified to give birth to modern bacteria. In the second stage—that of eukaryotes—several of these bacteria came together to form a chimera, modern eukaryotes: one bacterium gave cytoplasm; another, the nucleus; a third, mitochondria; a fourth, flagella; and so on. From the results of his work, which showed the existence of three types of ribosomes on planet Earth, Woese replaced that two-stage rocket with a three-branched tree, so that one branch led to bacteria, another to eukaryotes, and the third to archaea. All three emerged from the progenote (the name Woese gave to LUCA; figure 1.6). In this three-branched tree, modern eukaryotes no longer descend from prokaryotes in their current form, but from a line of prokaryotes that have disappeared; Woese called them urkaryotes. In this scenario, urkaryotes would have given birth to eukaryotes following two major events: first, the formation of the nucleus; then mitochondrial endosymbiosis.

I really liked the scenario Woese proposed because I had never agreed with the idea of a transformation of one (or several) bacteria into eukaryotes. Far from considering bacteria as primitive, I thought these organisms were perhaps the most elaborate products of evolution: miniaturized but extremely effective organisms that must have evolved from perhaps larger and more complex organisms—but less sophisticated, a bit like the superpowerful microcomputers of the 2000s have evolved through miniaturization from the first giant computers of the 1950s. Bacteria do indeed divide very quickly, and they rapidly renew all their macromolecules (this is called turnover). They also regulate the activity of their genes optimally by regrouping on their chromosomes the genes coding for proteins of related function: for example, the manufacture of an amino acid. This allows an easy regulation of this group of genes (called an operon) to adapt their expression to the needs of the cell.[2] Finally, the genes of bacteria are formed of a single piece: the information determining the synthesis of a protein is continuous from the beginning to the end of the gene. This is logical and straightforward, but surprisingly, this is not the case for our genes, those of eukaryotes. It had been discovered at the end of the 1970s that our genes are in pieces; the regions containing the information determining the synthesis of a protein (called exons) are interrupted by many "noncoding" regions that contain no information (introns; figure 1.3).

Eukaryotic genes are thus formed by a series of exons and introns. The translation of the message into a protein must be preceded by the cutting of the exons and their grouping in the right order to obtain at the messenger RNA level a coding sequence in a single piece, like a prokaryotic gene. This process is called splicing, a term first used by sailors to describe methods of joining two ends of a rope without making a knot. The splicing of the primary RNA transcribed from a eukaryotic gene (called primary transcript) into a messenger RNA is achieved in the cell nucleus by a large complex called a spliceosome. This complex, which is formed by many proteins and several small RNA molecules, resembles the structure of a ribosome, but is even larger. Some authors, such as the Nobel laureate Walter Gilbert, father of the term "RNA world," thought that this organization of genes into coding and noncoding regions was very ancient.[3] In his mind, exons corresponded to primitive coding genes for small proteins that had appeared by chance in the RNA sequences of the first organisms. During evolution, some of them would have succeeded in manufacturing larger proteins by inventing a complicated mechanism—the ancestor of modern spliceosomes—to connect the information of the small genes and produce larger messenger RNA. Eukaryotes would have preserved this primitive system of genes in pieces, whereas bacteria would have managed to get around it by eliminating noncoding sequences (introns) to possess genes in a single piece.

DNA Gyrase: The Magic Enzyme of Bacteria

I really liked this theory, as it agreed with my view of bacteria, which I refused to consider as inferior organisms. I was in the minority; at that time many molecular biologists, after studying model bacteria, abandoned them without a second thought to go on to study eukaryotes such as yeast, flies, or mice. If I considered bacteria to be highly evolved, it was perhaps because I was working at the time on a bacterial enzyme, DNA gyrase, which wasn't found in eukaryotes.[4] This enzyme possesses the extraordinary ability to activate circular DNA molecules by making the axis of the double helix turn on itself (gyration). This then forms a superhelix, which is called negative, because it turns in the direction opposite to that of the double helix.

It is this magic trick that facilitates in bacteria the separation of the two DNA strands for all processes that necessitate this separation, in particular the reading of the genetic message (transcription of DNA into RNA) and its reproduction (DNA replication). The negative supercoiling produced by the gyrase, which corresponds in a sense to an unscrewing of DNA, also modifies the three-dimensional structure of the genetic message. This influences the effectiveness with which genes will be transcribed into messenger RNA, then into proteins. Not all genes react in the same way to the unscrewing of DNA; this depends on the DNA sequence in the region that controls the activity of the gene in question. Some genes are activated by the unscrewing provoked by gyrase; others will be inhibited. The greater or lesser activity of the gyrase will thus suddenly influence the activity of all bacterial genes. The gyrase needs energy to form the DNA superhelices. It obtains this energy in the form of ATP (the molecule that constitutes the energy bargaining chip of all cells). Its activity thus depends on the amount of energy in the bacteria.[5] If the bacterium is in good form (rich in energy), the gyrase works perfectly and DNA will be overactivated, which will facilitate the expression of certain genes—those the bacterium needs in optimal conditions. If the cell is weakened (poor in energy resources), other genes will be activated—those the bacterium needs on a low-speed diet.

By playing with the activity of the gyrase, bacteria can thus instantaneously manufacture the range of proteins corresponding to their state of energy, which explains why these microbes can adapt very easily to changes in the environment: for example, from a relatively hot environment (such as our intestines) to a colder environment (I'll let you guess). But we, poor eukaryotes, don't possess that enzyme. How, in those conditions, can we consider ourselves superior? How could bacteria have lost such a fabulous enzyme as the gyrase by transforming themselves into eukaryotes? That seemed impossible. To me, the "urkaryote" ancestors of the eukaryotes Woese was talking about must not have resembled current bacteria, but "protoeukaryotes" possessing more primitive characteristics than those of bacteria, perhaps directly inherited from the progenote (for example, the absence of gyrase or even genes in pieces and spliceosomes).

I had always been interested in evolution, which wasn't the case among many of my molecular biologist colleagues. If I hadn't become a biologist, I might have been a historian and studied the history of civilizations. I wanted to understand the history of life and was fascinated by the existence of several large types of organisms on Earth—first prokaryote/eukaryote, and then the arrival of archaebacteria that shook things up. I hadn't studied evolution at school, but I had discovered its importance in biology by reading the Russian biologist Alexander Oparin's foundational work on the origins of life, which I came across by chance in the university library.[6] He explained well why it was impossible to understand biological subjects if one didn't know their history. He used as a metaphor the history of an extraterrestrial discovering on Earth a watch whose face is divided into twelve equal parts. The extraterrestrial, a genius by definition because he had been able to reach our planet, easily understood how this watch functioned, but was incapable of understanding why it had been divided into twelve. For that, it was necessary to go back to the history of the Babylonians, necessarily unknown to our visitor from the stars.

Reading Woese's article in *Pour la Science* was for me a true revelation. I felt I had to work on archaea (which were still called archaebacteria at the time). If, as Woese asserted, they represented a new group of living beings on Earth, it was easy to imagine that by studying these microbes we would only risk making new discoveries. At the time, I was a young researcher at the Institut Jacques Monod on the Jussieu campus in Paris, a stone's throw from the Latin Quarter, and I wanted to make my mark. Above all, I had a specific question I wanted to answer: Did archaea, like bacteria, possess a gyrase, or is that enzyme absent in archaea, as it is in eukaryotes? Because archaea are prokaryotes, since they don't have a nucleus, the presence of a gyrase in archaea would have suggested a link between this enzyme and the prokaryote phenotype. By contrast, its absence could represent a supplementary eukaryote characteristic of archaea.

I sought out all the publications concerning archaebacteria to see which ones I could culture. I couldn't work on methanogens, because I didn't have equipment that would allow me to cultivate anaerobic organisms (living in the absence of oxygen). However, I happily learned of the existence of aerobic archaea, which would be easy to culture: halophile archaea (lovers of salt) and the famous *Sulfolobus*, a thermoacidophile. I was thus amazed to discover the existence of microbes from hell: the beginning of a long love story. I read remarkable papers that Zillig had published on the RNA polymerase of archaea. And so I contacted him—with great respect—to ask for his help. He immediately agreed, sending me in the mail a lyophilized strain of *Sulfolobus acidocaldarius*. I was able to culture this strain by placing the receptacle (an Erlenmeyer flask), in which I would culture *Sulfolobus*, on a magnetic heating agitator (figure 1.5e). At the same time, I began studying halophiles, which were also aerobic and easy to culture; one had only to place a large quantity of sodium chloride (close to saturation) in their culture medium. I had obtained several strains of halophiles from an English scientist, Bill Grant, during a trip to Leicester to collect antibodies against gyrase from another British scientist. Germany (Zillig), England, France—the Europe of the archaea was doing well. I also brought along on the adventure a student, Christiane Elie, to study the replication of DNA in archaea. With halophiles, we obtained our first results and publication on archaea.[7] We proved the inhibition of DNA synthesis in archaea by aphidicolin, a drug known up to then to specifically inhibit eukaryotic DNA polymerase involved in the replication of chromosomes. These results confirmed those obtained by Zillig on RNA polymerase, revealing a new "eukaryotic" character in archaea. With *Sulfolobus* I would be able to answer my initial question: Do archaea possess a gyrase?

The gyrase belongs to a family of enzymes, DNA topoisomerases, which are present in all living beings and have enabled the resolution of problems posed by the winding of the two strands of the double helix, as we will see in the next chapter.[8] DNA topoisomerases are all capable of removing the negative supercoiling of a DNA molecule, with the exception of the gyrase that creates that supercoiling. Thus it was necessary to purify the DNA topoisomerases of archaea and see if one of them possessed the sought-after gyrase activity. This work was difficult to carry out in halophiles, because the proteins of these organisms need large quantities of salt to remain stable once they are extracted from the cell; this makes their purification very challenging, especially with proteins that interact with DNA. By contrast, the proteins of thermophiles, including their DNA topoisomerases, are very stable at low temperatures (we will see why in the next chapter) and thus are much easier to purify in the laboratory. I managed to convince a colleague in Paris, Michel Duguet, a specialist in DNA topoisomerase who was already a professor, to join me in the quest for the DNA topoisomerases of *Sulfolobus*. We rapidly formed a "task force" dedicated to this project along with two particularly motivated students, Gilles Mirambeau and Marc Nadal.

The results of our initial work on the DNA topoisomerases of archaea (and the results of a competing Japanese team) were completely unexpected; not only did *Sulfolobus* not possess a gyrase (like eukaryotes), but it possessed an enzyme that up to then had remained unknown—the reverse gyrase. Similar to gyrase, this enzyme introduces supercoiling in circular DNA molecules when mixed with these molecules in test tubes. However, the superhelices introduced by reverse gyrase turned in the opposite direction to those produced by gyrase. They are positive instead of negative (figure 2.1).[9] I will return to this story in chapters 3 and 5, because the reverse gyrase, which has since

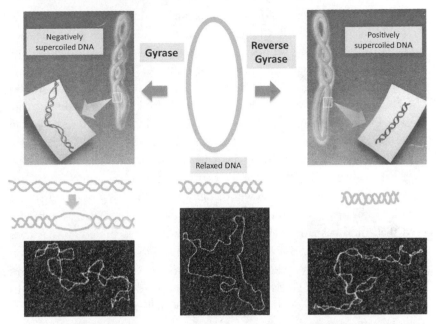

FIGURE 2.1. Gyrase and reverse gyrase. Left panel: negatively supercoiled DNA produced by DNA gyrase. The double helix is underwound and the two DNA strands can separate rather easily, producing DNA "bubbles." Central panel: relaxed circular DNA. Right panel: positively supercoiled DNA produced by reverse gyrase. The double helix is overwound and the two DNA strands are difficult to separate. The three pictures below represent plasmids DNA observed by electron microscope: two supercoiled DNA (*left and right*) and relaxed DNA (*center*). Courtesy of Eric Lecam, CNRS, France.

been found in all microbes from hell—and only in them—contains one of the main secrets of hyperthermophiles. We will see that there are still many mysteries to be solved surrounding this enzyme. For the moment, it's enough to know that my work on archaea, and in particular on the reverse gyrase, allowed me to quickly make a leap in the university hierarchy. I could finally defend my thesis in 1985, and in 1989 was given a position as professor at the Université d'Orsay, in a suburb south of Paris, with the supervision of a research team—all this for having read a popular scientific article during my vacation, lying in a hammock.

And so at the Institut de Génétique et Microbiologie de l'Université d'Orsay I set up the first French laboratory devoted entirely to archaea, and specifically to extremophile archaea, called Biologie Moléculaire du Gène chez les Extrêmophiles (Molecular Biology of Gene in Extremophiles), or BMGE. A long story full of developments thus began for me. One development was meeting my wife, Évelyne Marguet, a technician at the Centre national de la recherche scientifique (National Centre for Scientific Research), or CNRS, who had joined our team shortly after my arrival at Orsay. I mention Évelyne because we are now going to follow her on an exciting adventure she had in 1999 off the coast of Mexico, diving 2,600 meters under the sea in the submersible *Nautile*, looking for microbes from hell that live at those depths. (She went instead of me because I'm a bit claustrophobic.) This sample-collecting expedition was organized by CNRS and Ifremer (Institut Français de Recherche et d'Exploitation de la Mer). Zillig had also been invited on this trip. He was there as the number one expert in infernal viruses. Our goal was to inaugurate a new line of research focusing on viruses that infect anaerobic hyperthermophilic archaea living deep beneath the sea.

I will then describe the work we carried out with the samples brought back from this expedition. You will be plunged into a new world that is very popular right now—the world of vesicles. In looking for viruses, we discovered that microbes from hell produce vesicles that resemble viral particles (virions), but are not virions. We will see that some of these vesicles contain genomes of viruses. We will also see how our BMGE lab was cloned to set up a research unit on viruses of archaea at the Institut Pasteur in Paris. In the meantime, I'll leave you with Évelyne, who will describe her adventures 20,000 leagues under the sea.

June 5, 1999: AMISTAD Expedition (East Pacific Rise, 13° N, 104° W)—Évelyne's Onboard Journal

I slept badly . . . Today is the big day. In a few hours I'll be at the bottom of the ocean with more than 2,000 meters of water over my head: I'm incredibly excited. Impossible to sleep. When I woke up this morning, I thought of Daniel Prieur; he's the reason I'm here. Daniel, the director of research at CNRS, then professor at the University of Brest, was the first man in France to follow in the footsteps of Stetter and Zillig by searching for new hyperthermophile microbes. This man from Brittany, a lover of old ships, was of course immediately interested in microbes from deep below the ocean. At the end of the 1980s, he managed to convince the scientists at Ifremer to include microbiology in their research programs; since then, his team has participated regularly in expeditions Ifremer has conducted in all the oceans of the world. We met at the beginning of the 1990s, when Daniel and Patrick began to collaborate on the study of the plasmids (small circular DNA) of anaerobic hyperthermophile archaea to see in which direction the DNA superhelices turn in these microbes from hell. (Patrick has already talked about these superhelices and gyrases that turn in reverse.) Daniel found out that I was a fan of deep-sea diving. In 1997 he sent me to collect samples from hot springs at the bottom of Lake Tangankia, at a depth of 50 meters, for his research team. Today, he is sending me to discover a world that very few people will ever have the opportunity to observe with their own eyes.

Last evening, along with all the other scientists on board the Ifremer ship (the *Atalante*), I attended the briefing given by Christian Jeanthon, the head of the mission. It's the same ritual every evening. All the participants in the AMISTAD expedition (Advanced Microbiological Studies on Thermophiles: Adaptations and Diversity—*amistad* means "friendship" in Spanish)—gather to prepare the following day's dive. We are fifteen scientists of six different nationalities participating in this international expedition in the eastern Pacific, off the coast of Mexico. We will each be diving once: you can't miss your turn. There's no question of being sick on D-day; there is no second chance. And so that evening, I listened to Christian's speech with even more concentration than during earlier briefings. He told us the site where we would be diving the following day, the types of samples we were to bring back, and the various maneuvers we were to carry out. We were also to bring back the equipment left behind during earlier dives. With each dive, the scientist sent down is not just working for himself or herself; he or she also participates in the experiments of colleagues who have gone down before or who will go down next. I paid a lot of attention to all that I was to do during the dive: noting everything I could observe so I could prepare a detailed report on the dive, and take photos of the different sites we would visit and the equipment we would place on the smokers. At the end of the briefing, I examined the map showing the location of the principal active hydrothermal sites we were going to explore. They are a two-day sail from Manzanillo, the Mexican port we departed from, on a fault (called East wrinkle) of the great ridge that surrounds the Pacific Ocean. On this map I deciphered names that later became familiar: Grand Bonum, Pulsar, Genesis, the Chain, and so on. It was there, in those fields of undersea smokers, that I would navigate in a few hours, at a depth of exactly 2,630 meters.

Claire Geslin, my roommate, is a doctoral student in Daniel Prieur's lab. She's still sleeping peacefully. She will no doubt be less relaxed the day of her own dive. Outside, the sun is already shining and throwing rays of light on the walls of our cabin. After breakfast, it's the great dive. I refrain from having tea because there aren't any toilets in the submersible. Finally, at 9:30, we get started. The first important thing is to put on the suit to protect against the cold I'll soon find in the depths. The suit is yellow like our little submarine, the *Nautile* . . . a yellow submarine (figure 2.2a).

Putting on this suit is no easy feat. You can't keep from thinking about the astronauts . . . OK, I'm all set. There's no turning back now.

The *Nautile* is there, waiting on the deck of the oceanographic ship. This

FIGURE 2.2. The AMISTAD expedition. *A*, the submersible *Nautile* from Ifremer (Institut Français de Recherche et d'Exploitation de la Mer). *B*, Évelyne and the copilot. *C*, a white smoker and the *Nautile* arm ready to collect samples. *D*, a fragment of chimney. *E*, Zillig working in the anaerobic chamber of the *Atalante*. *A*, *B*, *D*, *E* courtesy of Évelyne Marguet (Université Paris-Saclay). *C* courtesy of Ifremer.

dive will be its 1398th. Enough to reassure the neophyte: Why would today be the day when it encounters its first serious problem? There is no reason at all. Of course, there's always a first time. A small twinge, a rise in adrenaline . . . not so bad. I reassure myself: the pilot and copilot, Jean-Paul and Yann, are inside the *Nautile* carrying out the final checks. A few days ago they went over the submersible's entire electrical system with a fine-tooth comb to repair a minor technical issue. It was impressive, all those cables spread out on the deck of the ship. I hope they put them back in the right places. Stuffed into my yellow suit, I wait for what comes next. All the other participants in the AMISTAD expedition and the sailors of the *Atalante* are there. They're watching me. Every day, the *Nautile*'s departure is a great occasion that everyone wants to witness. Today, I'm the heroine of the day. The pilot and copilot have emerged from the airlock chamber. All is OK . . . I can proceed. I'm the last to go in, putting one leg after the other through the narrow opening on top of the submersible. I then go down a ladder and then lie down on a mattress.

There are three of us. The pilot (Jean-Paul) and I are on two parallel bunks, each facing a porthole (figure 2.2b). Yann, the copilot, is sitting on a seat a

bit above our bunks, behind the controls. The interior of the submarine resembles an airplane cockpit, but much smaller, of course. So many levers. So
many buttons. So many recording screens. Soon I hear the airlock door closing on us. This time it's serious: the three of us are gathered here for better or
for worse. The pilot and copilot give me the list of instructions to follow in the
event they are incapable of carrying them out. What exactly does that mean?
Why do I need to know what to do in case of fire, depressurization, or lack
of oxygen? Hey, guys, are you joking, or what? I hope I won't have to use this
new knowledge . . . I feel another burst of adrenaline.

I scarcely have time to digest all this when the *Nautile* begins to move.
It first glides on rails to the edge of the ship's deck. The movement is slow. I
feel like I'm in a film in slow motion. Through the porthole I see my friends
and the sailors who are watching us leave. I recognize Wolfram Zillig, who
towers over the rest of the group. He is supposed to go down in the *Nautile* in
a few days. The Germans don't have a submarine like this, and that pioneer of
research in microbes from hell hasn't yet had the opportunity to see with his
own eyes the hydrothermal chimney of an undersea smoker. I know he would
really like to be in my position right now.

I see the heads of my expedition friends go by: Claire, Christian, Nicolas,
Viggo, Max, Nathalie, Edmond, Bernard, Marianne . . . They're all there, then
they disappear: we're leaving the "sailors' diving board." Suspended from a
movable crane by a thick central cable and ropes on the sides, the *Nautile*
first rises into the air (a curious sensation for a submarine) to reach its diving position. We are now suspended between the sky and the sea, above the
waves of the Pacific. Slowly, the *Nautile* is placed gently on the surface of the
ocean. It floats thanks to several reservoirs filled with air, ballast, located on
either side of the submarine shell. Two divers have jumped into the water to
remove the four ropes that still attach us to the *Atalante*. We can see one of
them through the porthole; he gives us the thumbs up. As a diver myself, I can
imagine their feelings. The pilot is speaking on the radio with those on the
ship responsible for the dive: everything is in order; we can begin our descent.
For that, the air contained in the ballasts must be replaced with compressed
air. The *Nautile* will then go straight down, weighted down by more than a ton
of iron gravel. We are to arrive in the zone of the smokers, because thanks to
a system of Argos markers, the *Atalante* is positioned directly above the site
to be explored. These markers were placed by geologists in the 1980s, during
the *Nautile*'s first dives.

Suddenly we're in the great blue sea and, very quickly, in total darkness.
Even for me, someone who has done a lot of diving, it is a shock. I know what
it's like to be under the ocean (up to 50 meters below, which isn't bad), but

total darkness . . . We will descend for an hour into this completely empty universe. Very quickly I'm overcome by an extraordinary feeling of excitement: −60 meters, my diving record has been broken; −100 meters; −200 meters . . . amazing. Fortunately, the cabin is pressurized. Our descent began in great silence, but very soon my companions start to talk. Lots of joking. We're now lying on our backs listening to music from a DVD player. We doze . . . a bit worried, very excited, as the wait continues.

We've been descending for more than an hour. We're getting close to −2,000 meters. The temperature inside the *Nautile* is now 10°C. Luckily, our suits keep us warm . . . We're now approaching the bottom of the ocean. The excitement grows. Our eyes are riveted to the expanse of the abyss revealed by the lights of the projectors. Suddenly, we're there! After an hour and a quarter of descent we touch bottom. The ground appears suddenly: moon rocks and sand at 2,630 meters of depth. It's incredible. What a sensation! My heart is pounding . . . the pilot releases 250 kilos of iron gravel to stabilize the *Nautile* just above the ocean floor. He then starts the small motor that will give us the freedom we need to move around in search of the site where we are to collect samples. We're looking for the first of three sites we are to visit during this dive; the code name of one site is PP55 (PP = *petit poucet*: "Tom Thumb"). We don't doubt for a moment that we will quickly reach our goal. But after a quarter of an hour, a muted worry begins to settle in. We know that the smokers we want to find should be within a radius of a few kilometers of our marker. But in the utter darkness of the ocean floor, the thin ray of our projected light forms only a small circular zone in front of the submarine. There is a risk that we will pass next to a site without seeing it. A few days ago, during an earlier *Nautile* dive, the pilot was unable to find the zone of the smokers that had been assigned to him. After wandering around for several hours on the ocean floor, which is dishearteningly flat and uniform, the crew had to return empty-handed. What a disappointment: an entire day of work lost, not to mention the cost of a *Nautile* dive. Are we going to have the same bad luck? Anxiety begins to take over; everyone is silent. Finally, after twenty minutes of stress, a hidden cathedral rises up before our amazed eyes. The video cameras begin to roll. Silence: we're filming. We navigate above a forest of extinguished smokers (also called vents), but there are so many chimneys that they form the landscape of a buried city, an Atlantis of rocks, arches, columns, and temples. Some chimneys measure several dozen meters high. They have grown while taking on fantastic shapes, giant stalagmites formed by the minerals dissolved in heated water in the bowels of Earth that then precipitated on contact with the icy water of the ocean (figure 2.2c).

Careful, the water is moving—it vibrates . . . We have arrived at the active zone PP55.

I take a few photos with the *Nautile*'s exterior camera to immortalize this moment. The first animal form appears attached to the small smokers; the decor is illuminated with the colors of life . . . oases at the bottom of the oceans. It takes my breath away. We advance very slowly, and very rapidly we are plunged into fairyland. The rocky bottom is covered with sea anemones; I see many Galatheas, as well as white crabs with elongated bodies and long legs, which are climbing on the walls of the small chimneys scattered around the site. Long, white fish with bizarre heads (crab fish) slither among them. These wonderful hydrothermal vents are true works of art; one of them, ocher in color, is carved in an astonishing way.

We finally reach our first goal, a large chimney called Pulsar. Its upper cone, out of which escapes a plume of black smoke, is covered with Alvinella, small worms (annelids) that live in colonies and build organic tubes fixed to the walls of the stacks. They live inside the tubes, whose base is in direct contact with burning fluid. They are the most thermophile animals known; they can survive for a fairly long time at temperatures close to 50°C.[10] Their head, which resembles a small red star, comes out in contact with the colder water that surrounds the top of the tubes. Soon I notice at the base of Pulsar a small break on which a metal basket and a temperature gage had been placed a few days earlier, which we are now to recover. The basket is a microbe trap. If bacteria or archaea are transported by the mineral particles carried by the fluid that comes out of the break, they will be caught in the metallic mesh inside the basket. This basket belongs to Max Mergeay, a Belgian microbiologist who is studying microbes' resistance to heavy metals. Underwater smokers emit a large quantity of these metals, which then accumulate at their base. Max wants to know whether these sites are little paradises for bacteria resistant to heavy metals. He usually goes looking for these bacteria on polluted industrial sites. He is all the more grateful to be participating in this Pacific expedition with us. I note the temperature indicated by the probe that had been placed under Max's basket: 25°C . . . if bacteria live here, they won't be hyperthermophiles. Jean-Paul manipulates one of the two remotely guided arms of the *Nautile* that ends with a pincer (figure 2.2c), which enables him to recover the basket and the probe. He then places these items inside the exterior container located in front of the *Nautile*. I am amazed at how deftly he carries out these operations.

Once the basket has been retrieved, we need to find an adequate location in which to place two other microbe traps prepared by an Icelandic scientist,

Viggo Marteinsson. He studies the effect of pressure on hyperthermophiles. During this expedition he hopes to find microbes whose growth is stimulated at high pressure (piezophiles, from the Greek *piezo*, "to press"). Viggo's metallic baskets will stay there for a few days before being retrieved by another team. To be sure that his microbe traps will be found easily in the smoker environment, he placed a red plastic cube, on which he has drawn a smiley face, on top of them. After several temperature readings at the fissures located on a platform at the base of Pulsar, we find a relatively flat site whose temperature is 70°C, which seems right for this experiment. The pilot moves the *Nautile*'s arm to place the two baskets on the platform one after the other; everything goes smoothly. I take photos and a video of the site with the baskets so I can show them to Viggo when we return.[11]

After forty minutes of uninterrupted work, we leave Pulsar and go to new vents, PP71 and PP57, where we are to carry out the day's experiments. Here we are again wandering above an "arid" desert landscape. For a moment, we hover above a field of dead chimneys. The reliefs are superb. The average lifespan of a smoker is a dozen or so years; after a certain amount of time, the movement of the ocean floor causes the paths of the fluids that move to the surface to vary. Old stacks are extinguished; new ones light up. What happens to our favorite microbes when a black smoker dies? Once again the famous Karl Stetter, whom Patrick has mentioned, found the answer. During a research expedition also carried out in the Pacific Ocean, the boat that sheltered his expedition was almost carried away by the eruption of an underwater volcano called Seamont in the Polynesian archipelago. Fortunately, the ship was still a two-day sail away from the site of the eruption. Karl then had the idea of taking, at a depth of 200 meters, a sample of water containing a bit of the vapor cloud, mud, and rocks emitted by the volcano. After being brought back onto the boat, this liquid, whose temperature was that of the ocean (2°C), was incubated at 80°C in a culture medium for hyperthermophiles. A few days later, an entire world of infernal microbes—*Archaeoglobus*, *Pyrococcus*, and *Thermococcus*—were thriving in the culture medium. The hyperthermophiles had been spit out by the volcano and had floated in the icy water, slowed down by the cold, until the miraculous moment when they found themselves on a boat chartered by some slightly crazy microbiologists who had brought them back to life by simply putting them back in their preferred temperature, where they once again frenetically divided. Stetter and his team had a new scoop: another well-deserved publication in the journal *Nature*.[12] Their observation showed that hyperthermophiles could easily survive a prolonged swim in the icy waters of the ocean. It is this capacity of microbes from hell to survive at low temperatures that is the guarantee of their dispersal. For them, each volcanic eruption must be a gift, the occasion to see new sights. Speaking of volcanic eruptions, I hope that one of the volcanoes that is sleeping no doubt not very far from here doesn't choose today to wake up.

It's been an hour since we left Pulsar and we're still looking for PP71. Yet on the map, that site seemed very close to our point of departure. Because I have been staring at the space in front of me being swept by the rays of the *Nautile*, my eyes are beginning to hurt. My stomach starts to remind me that it might be time to relax a bit and take a lunch break. My dive companions appear to agree, so we take out our meals from their drawers and have a well-deserved snack. Then we start off again. We decide to stop searching for PP71 and focus on PP57. This time we're lucky. After ten minutes or so, we discover the new site (figure 2.3). It is even more spectacular than the first. I'm holding my breath and my heart is pounding; it is more beautiful than anything I've ever seen in photos or on film. The smokers are surrounded by a multitude of giant worms with red plumes, the famous Riftia, characteristic of underwater

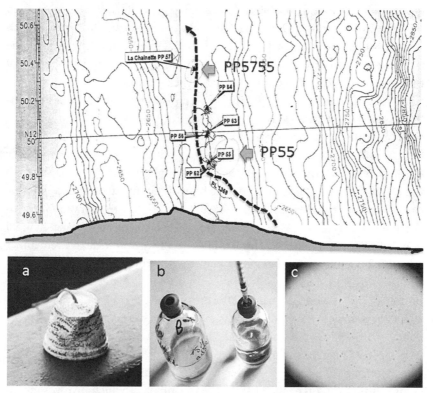

FIGURE 2.3. Map of the East Pacific Rise explored during Évelyne's dive (*dotted line*). A, effect of pressure on a polystyrene cup attached to the *Nautile*. B, Penny vials. C, optical electron micrograph of *Thermococcus* cells isolated from AMISTAD samples.

oases. These animals of the abyss form a living tapestry out of which emerge plumes of black smoke. In the midst of this forest of Riftia, I finally notice crabs and a few shrimp. The site is teeming with life.

We arrive at the foot of a huge black smoker called the Chain ("La Chainette" in French). Its walls are covered with white filaments, layers of bacteria that feed on the gas emitted by the breaks located on the sides. We look for a place to put two microbe traps made by Christian Jeanthon, the head of the expedition. We are to place the baskets on breaks out of which a high-temperature liquid, between 70 and 90°C, escapes. Another team will recover them in a few days. Christian's traps are unique in that they are filled with fiberglass. He hopes to trap a larger number of microbes this way. They will be placed on several sites during the AMISTAD expedition to determine whether the vent species of archaea and bacteria are found everywhere, or vary from one vent to another.[13]

The pilot maneuvers the *Nautile*'s arm and delicately places Christian's two baskets on a ledge in the middle of a small smoker. Everything seems to be going smoothly. But then a big crab, which is not at all happy to cooperate in the experiment of the day, appears. He doesn't want to share his territory with these intruders from the surface. He abruptly picks up, one after the other, the two baskets with one of his powerful pincers and throws them to the bottom of the smokestack. The pilot responds, again placing the baskets in position, and again there is the war of pincers. The crab's pincer versus the *Nautile*'s pincer; the advantage is not necessarily on our side, since the crab is much more agile than our little submarine. Christian should have prepared much heavier baskets. When we put one basket back in place, our crab takes the opportunity to knock over the other one. Our crustacean won't give up. We push him with the *Nautile* arm to make him fall to the base of the smokestack. Success—we can now put our baskets back in place. But scarcely is the operation finished when the crab is back. He has crawled up and again sends our basket flying. I wonder if we have really upset him or he is simply toying with us. I remember a lecture by an animal behavior specialist who explained that animals love to play. This little game thus continues for a while: we push the crab to the bottom of the smokestack to install the baskets; he climbs back up and throws the baskets down. After four fruitless attempts on our part, the situation becomes more serious: our crab is joined by his colleagues, who seem to want to help him and take part in this underwater handball match. Confronted with such determination from the animal world, we decide to abandon the game.

It is time to go on to the final part of our mission, which involves me directly. We are to take a piece of a smokestack and bring it back to the *Atalante*. With each dive, samples of smokers are thus collected and then distributed to participants of the AMISTAD expedition. I will hunt for viruses in these smokestack fragments. The pilot is still moving the submarine while the copilot and I are dealing with measurements. I determine the temperature of the external wall of a smokestack that is emitting particularly thick black smoke. The temperature of the fluid coming out of the breaks on the wall of this chimney is 60°C: it's not particularly hot. The probe now approaches the fluid coming out of the mouth of the smoker: 300°C . . . 310°C . . . the thermometer stops at 350°C. Of course, we know that life can't exist at 350°C, because at that temperature the agitation of atoms is such that proteins and nucleic acids are destroyed almost instantaneously. But between the fluid at 350°C and the external surface of the chimney at 60°C there is a temperature gradient in the wall that must enable the development of our microbes at around 100°C. So we must remove fragments from this wall. There's only one way to do that: we must break off a piece of the chimney by using the *Nautile*'s arm and have the rock fragments fall directly into the exterior case in front of the submarine. The operation is a complete success, and we recover a fragment from the top of our chimney.

After five and a half hours on the bottom, we finally receive permission to go back up. The *Nautile* descended like a rock but it will go up like a bubble. It is extremely easy for it to regain the surface: the pilot only has to release the rest of the iron gravel in the containers on each side of the submarine. We quickly rise above the chimneys of the Chain; at 50 meters above the ocean floor, we are still at the top of some of them, which gives us a good idea of their height. The *Nautile* gradually rises, and I try to recall what I have just experienced. Everything went very well. I am in a slightly euphoric state. We have just seen one of the most fascinating parts of the planet—I am well aware that I am one of the privileged few to have witnessed it. The five hours spent on the bottom went by very fast. The climb will take an hour. Finally, our journey ends and we emerge at the surface of the ocean. The pilot replaces the water in the ballasts with air to enable us to float. The ocean is rather calm today, which is a gift. I remember less lucky predecessors who, during another expedition, had to stay eight hours on the surface during a storm, bounced around wildly by waves, before the *Nautile* could be brought back up onto the deck of the *Atalante*.

The entire crew is back on the deck to witness the return of the little yellow submarine. Our colleagues are eager to recover the samples. The divers come to attach the cables that will pull us up onto the deck. They also place a small parachute behind the *Nautile* to stabilize it. This wind boom will allow it to stay at the back of the *Atalante* during the recovery phase. At that moment, I realize that my initiation ceremony will soon begin. Impossible to escape it:

this was my first dive, and it must be celebrated. Scarcely after being extracted from the *Nautile* hatch, and having removed my suit, I am led to the chosen spot under the teasing glances of colleagues who have already undergone this initiatory rite and the nervous looks of those who have not yet descended. I'll spare you the details; you only need to know that face decoration with melted chocolate is part of the ritual. And to wash it off, there's the fire hose. Once the initiation rite is over, I ask to see the eighteen polystyrene cups that Barbara, one of the Americans onboard, had strung outside the *Nautile* before the dive. Each member of the expedition had signed his or her name on one of the cups. She wanted to measure firsthand the effect of the pressure exerted at a depth of 2,600 meters (260 atmospheres) and have a souvenir for each of us. The results were impressive: although we could still identify our signatures, the cups were completely crushed (figure 2.3a).

How do our microbes from hell manage to resist such pressure? Such an ability might seem incomprehensible to most people. For a biologist, however, there is nothing surprising in it. We find many classic microbes, but also fish, at even greater depths. What makes us sensitive to differences in pressure are our lungs, whose constitution is similar to that of the air-filled polystyrene cups we brought down with the *Nautile*. Under the effect of pressure, the bubbles diminish in size, causing the breakdown of the entire structure of the receptacle. That is why the cabin of the *Nautile* must be pressurized.

Only a couple of hours after arriving back at the *Atalante*, here we all are, sitting around one of the large tables of one of the floating laboratories on the boat. The fragments of chimneys recovered today (figure 2.2d) are placed on a piece of aluminum foil. Seated in front of them, Christian, as head of the expedition, will preside over the division of the samples. He is the only one authorized to do this. He puts on gloves and, with a hammer, detaches small fragments of the smokestack, which he stingily hands out. One after the other, each scientist holds out a tube so as not to leave empty-handed. After the distribution of samples, it's mealtime. The food on board is excellent: the cooks are first-rate. We all have a thought for our American colleagues, who have only Coca-Cola and iced tea on their oceanographic vessels. On the *Atalante*, there's plenty of good wine. Everything is a reason to celebrate. In a few days, it will be the *Nautile*'s 1400th dive; a feast will be laid out on the deck and we will all receive a diploma certifying our dive. The ambience is warm. The sailors eat with us: they've decided to tease the few scientists who don't have a sense of humor. When we leave each other after three weeks of group living, many will have tears in their eyes.

Once dinner is done we return to work. There's no time to lose and the days are too short. During the evening debriefing I begin by giving a talk on my dive: it's the occasion to see the photos and videos taken during the day. It's now 11:00 p.m., but it's not yet time for bed. We must return to the lab and put some of our samples into a culture medium. Will the rock fragments recovered today "give birth" to new hyperthermophiles? Using a mortar, I crush the fragments carefully in sterile seawater. If they are there, the poor microbes must be really cold right now. Also, they must be scared of all those oxygen molecules in the room. When the microbes are cold, those molecules can't hurt them much. Our microbes from hell must from now on be in the presence of a crowd of other microbes that have fallen on our samples, because they have been carried to the surface without particular precautions. This means they have been contaminated by microorganisms that were in the air, on Christian's gloves, on the mortar, and so on. Luckily, all these contaminants are not able to live at 80 or 90°C. Soon they will burn in hell and we'll never hear from them again.

To try to grow in the *Atalante*'s lab the *Thermococcales* that up to then had lived in the wall of our underwater chimney, I will add my chimney powder soaking in sterile water to a nutritive liquid (called a culture medium). The choice of culture medium is very important, because it will determine the nature of the microbes that we can hope to grow. In the present case, I want to obtain hyperthermophilic archaea of the order *Thermococcales* (*Thermococcus* or *Pyrococcus*). *Thermococcales* need organic molecules to grow, because they are heterotrophs (see box 2.1), and so our culture medium must contain a mixture of complex organic molecules (in particular, proteins). We decided on yeast extract (from microscopic mushrooms), a classic of the genre. The medium must also contain mineral salts, including certain rare metals, and flowers of sulfur (S). *Thermococcales* use this element to eliminate protons (H^+) produced by their breathing in the form of hydrogen sulfide (H_2S; thus the smell of rotten egg that fills the lab if a clumsy scientist spills or breaks a bottle containing a culture of these infernal microbes). We call this medium the Wolfram (Zillig) medium, because he's the one who concocted the recipe a few years ago.

I take a small bottle, called a Penny vial (from the name of a scientist; figure 2.3b), half filled with the Wolfram medium, and add my solution of liquid smoker powder using a pipette. I close the vial with a rubber stopper, then place on the stopper a metal capsule cut in the middle that I seal with crimping pliers onto the neck of the vial. This capsule seals the vial hermetically and will prevent the stopper from coming out; if it came out, the hydrogen sulfide produced during the growth of our little friends would be thrown into the air of the room. The circular opening in the middle of the capsule enables a needle to pass through the rubber stopper. Once the needle is taken out,

BOX 2.1. RESTORED TO HELL

Microbes can feed in many ways, and microbes from hell are no exception. Some of them use organic molecules already constructed by other living beings to produce their energy and manufacture their own macromolecules. These are heterotrophs (like us, by the way), which need to eat other living beings (or at least parts of them) to live. Heterotroph hyperthermophiles that thrive in underwater oases have an embarrassment of choices: fauna that is visible (worms, mussels, and other giant crabs) and invisible (bacteria) that gravitate around the smokers, which are a rich source of such molecules. Other hyperthermophiles are capable of feeding on and "breathing" gases and minerals that escape from the smokers. They are, in a sense, eaters of stone and drinkers of gas. They derive their energy from the hydrogen that is abundantly present in volcanic vapors. They can manufacture their own organic molecules by using the carbon present in the carbonic gas that escapes from the depths of the earth. In scientific terms, these are autotrophs: they are completely autonomous for their breathing and feeding. This ability is found not only in microbes from hell: it is widespread among bacteria and archaea, but we find it also among eukaryotes such as green algae and plants. Still, the latter need energy from the sun to synthesize their organic molecules from atmospheric carbonic gas—this is *photo*synthesis. Many microbes, however, use only the energy from chemical reactions, causing hydrogen to intervene to manufacture their organic molecules from mineral matter—this is *chemo*synthesis.

the hole closes instantaneously. We can thus take out or add gas to the vial. Initially we must remove all oxygen in the vial, as it is highly toxic for *Thermococcales* when they are grown at a high temperature. I begin by piercing the stopper of the vial with a needle attached to a syringe (figure 2.3b). This syringe is connected by a rubber tube to a pump that allows me to aspirate the air inside the vial (thereby emptying it). Then, using the same procedure, I fill the vial with nitrogen from a pressurized bottle of gas kept in a corner of our lab. This dual operation, which scientists call empty-gas, must be repeated several times to be sure that no oxygen remains inside the vial. The culture medium, which was pink at the beginning, becomes colorless after this series of empty-gas, which proves that all the oxygen has indeed disappeared. We observe this change in color thanks to a colorant, resazurin, which we added to the Wolfram medium. Everything is ready. We just have to place our vial in an incubator at 90°C (the optimal temperature at which *Thermococcales* grow) and hope . . .

Well, it's late. What a day it has been! I take a final look at tomorrow's sched-
ule: it will be Edmond Jolivet's turn (he's another doctoral student of Daniel
Prieur's) to go down with the *Nautile*. Edmond is studying the mechanisms
that enable some hyperthermophiles to tolerate doses of radiation that are
usually deadly for all other living beings (with the exception of a few radioresistant bacteria, such as the famous *Deinococcus radiodurans*).[14] This time I
will be on the deck of the *Atalante* to encourage our young colleague. It's time
to turn out the light . . . and have sweet dreams. This time I won't need to be
rocked to sleep. I've just lived a magical, unforgettable day.

June 7, 1999

Two days later: success! The culture medium, which was clear at the start,
has become cloudy. I place a drop of it on a glass plate to observe it under
the microscope. There they are. Little black spheres moving around in every
direction; no doubt about it, they are indeed our favorite hyperthermophile
archaea, the *Thermococcales* (thermal shells; figure 2.3c). They don't seem
happy. I placed them abruptly under an intense light, at a more than polar
temperature (for them) and in the presence of oxygen. Some cells were dividing; they appear in the form of small spherical doublets. Just forty-eight
hours ago they were calmly living their lives under two kilometers of water
in a sunken city and now they are on board an oceanographic vessel. Soon
they will be studied in our laboratory at Orsay at the Université de Paris-Sud.
Using the same process, we obtain several other "positive" cultures—that is,
containing hyperthermophile microbes (which we call "enriched cultures").
We want to isolate the greatest number of different strains of *Pyrococcus* and
Thermococcus to increase our chances of finding one (or several) infected
with viruses. We don't know of any viruses that "parasite" these anaerobic
hyperthermophiles: most viruses that have been isolated so far by Zillig attack
aerobic thermoacidophile archaea close to the genus *Sulfolobus* that are found
in acidic terrestrial hot springs. All the viruses Zillig has revealed have a property in common: they don't kill the cell that harbors them, but are content to
hide inside while reproducing at the same rhythm as the cell.[15] Some viruses
nonetheless escape from their host from time to time to go infect other cells.
The hyperthermophiles that harbor these viruses are thus healthy carriers, but
contagious (luckily not for us, as viruses that attack microbes from hell are
so well adapted to the conditions of life in their hosts that they can't multiply

at 37°C). With our new research program focusing on *Thermococcales* from deep-sea vents, we want to know if the viruses that infect anaerobic hyperthermophile archaea living at the bottom of the oceans resemble those that infect the aerobic hyperthermophile archaea of the terrestrial springs. Then we can compare them to known viruses that infect bacteria or eukaryotes. We hope to tackle fundamental questions: How long has the cohabitation—which sometimes turns to confrontation—of viruses and cells existed? Where do the viruses come from? We are attempting to answer this type of question by studying the viruses that we hope to isolate from our AMISTAD samples.

To successfully carry out this research project, we count greatly on Zillig's help, as he was the first person in the world to isolate viruses infecting aerobic thermophile archaea. The first positive result of our new research program was my participation in the AMISTAD expedition alongside this legendary scientist (figure 2.2e). I had already worked with his team in Munich to learn the basic techniques of virology at high temperatures, and we had shared a space in the lab of the *Atalante* during the entire expedition. We got along very well. He invited me to return to work with him in Germany to isolate new strains of *Thermococcales*.

Évelyne's onboard journal stops there. We will now jump ahead fifteen years to see what the AMISTAD expedition gave us. Unfortunately, we didn't have the opportunity to go back down in the *Nautile* after that expedition, but we had recovered enough material to fill the life of a scientist. In the past few years, Prieur has retired and the *Nautile* has often been replaced by a robot, affectionately called Victor, piloted by scientists from the deck of the *Atalante*. This is a very effective method, but it is much less exhilarating than the excursions of our first explorers of the abysses. To get back to Évelyne and the other protagonists in our story, let's now return to Paris, to a good meal—a very French way of celebrating science.

January 2014: Fifteen Years after AMISTAD

We are all gathered to celebrate the new year at La Crémaillère, a restaurant in Bourg La Reine in the suburb near Paris where Évelyne and I live. Most of the current members of the two BMGE teams are here: those from the Orsay lab and those from the BMGE unit of the Institut Pasteur. (We will see at the end of this chapter how archaea were able to take over the prestigious Institut.) We share an excellent meal prepared by one of the former chefs of the transatlantic ship *France* in this restaurant that serves traditional French cuisine at very reasonable prices (perfect for scientists, students, and postdoctoral fellows with low salaries). We are also celebrating the European Research Council (ERC) grant I have just obtained that will last for five years starting next month. I remember fondly the moment when I received the good news on my iPhone. It was in the Charles de Gaulle Airport, where I was waiting to board the plane that would take me to the United States to attend an international conference on archaea. ERC grants offer a sizable sum, 2.5 million euros (close to three million US dollars) for five years, which enables us to hire several young out-of-work scientists and thus increase their chances of one day finding a permanent position in research.

I look around at my tipsy colleagues sharing a Norwegian omelet, La Crémaillère's specialty; it was thanks to several of these colleagues that I was able to obtain this funding. The project I proposed, called EVOMOBIL, is in large part the direct consequence of the work carried out for fifteen years at Orsay on samples brought back from the AMISTAD expedition. Let's go back to the beginning, the year 2000, and see how research advances in a sometimes unforeseeable way. Let's go around the table and introduce the protagonists in this adventure. At the head of the table is Évelyne. Upon her return from the expedition, she spent two weeks in Munich studying the installation that Zillig had put in place in the 1990s to culture anaerobic hyperthermophiles so she could reproduce it at Orsay. This installation included an anaerobic room, where one could work in the absence of oxygen; jars that enable petri dishes to be placed under pressure at high temperatures; and a network to distribute various gases, some of which were deadly, or explosive, such as H_2S and hydrogen (similar equipment was also used on board the *Atalante* during the AMISTAD expedition; see figure 2.2c). To form a solid gel that would enable *Thermococcales* to grow on these dishes, we don't use agar, as Petri did in the first chapter, because it melts at 80°C. Instead, we use Gelrite, a natural polymer of sugar that, once it has solidified, does not melt at 100°C. We also go back to using the good old glass dishes, since plastic would melt at high temperatures.

Zillig's lab was filled with small instruments, one more ingenious than the last—in particular, a "bottle warmer" enabling a number of glass petri dishes to be maintained at the right temperature, 85°C, and to deliver them one by one, ready to be used. I joined Évelyne in Munich for a week to isolate new strains of *Thermococcales* that would, we hoped, enable us to discover new viruses. In a week we had isolated seventy pure strains that we could take back to Orsay.[16] Each strain came from a colony obtained from a single cell placed on a petri dish (figure 2.4). To do this, we had to dilute an enrichment culture obtained in a liquid medium and place it on a layer of colloidal sulfur covering the solidified Gelrite on the petri dish. The development of a

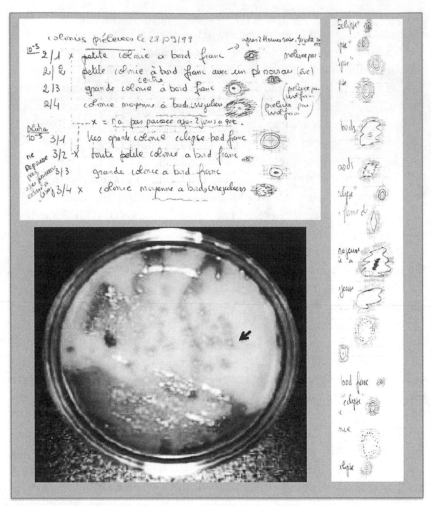

FIGURE 2.4. Extracts from Évelyne's notebook reporting the isolation of new *Thermococcus* species in Munich, with drawings of plaques formed by growing colonies on Gelrite plates. A picture of a petri dish covered with white colloidal sulphur in Gelrite. The arrow indicates a white colony of *Thermococcus* growing in the middle of a black zone corresponding to the dissolution of colloidal sulphur.

colony could be seen by the appearance of a light spot (a plaque) on the layer of yellowish sulfur. *Thermococcales* will dissolve the layer of colloidal sulfur during their growth by division from the initial cell. Colonies exhibit different shapes (see the handmade drawings in figure 2.4), indicating that each corresponds to a different *Thermococcus* strain, with different physiological properties. Initially, we had given our seventy strains numbers based on the number of the enrichment culture that had served to isolate them from the plaque from which they came: for example, 30–1 (one dish being able to show several plaques). Later, some of these strains would acquire a certain fame . . . at least within our team.[17]

Back at Orsay, Évelyne was able to establish a platform for culturing anaerobic hyperthermophiles as effective as Zillig's. But not without some difficulty. A location had to be found, as did funds to equip it with the right equipment: in particular, the anaerobic enclosure. It was also necessary to install a system to connect the containers of gas, located outdoors for safety reasons, with the hoses used in the anaerobic enclosure, for the jars and Penny vials in the lab. Some colleagues were not happy to see canisters of hydrogen (H_2) or hydrogen sulfide (H_2S) being unloaded near the buildings of the institute. Why go to so much trouble when we could have continued to work with aerobic hyperthermophile archaea, like *Sulfolobus*, which are much easier to manipulate? Yes, but the viruses of *Sulfolobus* were becoming well known, and their uncontested master was Prangishvili, who was working at the time with Zillig in Munich, so we had to confront new challenges. Isolating the viruses of *Thermococcus* was one of them. At the end of 2002 we were ready to set off in search of them.

Fifteen years later, we still haven't found a virus of *Thermococcales* at Orsay. Are we such bad virologists? Perhaps. In any case, it seems much more difficult to isolate viruses of anaerobic hyperthermophiles such as *Pyrococcus* or *Thermococcus* than to isolate viruses of *Sulfolobus*. Zillig himself tried in vain during a yearlong stay at Prieur's lab in Brest in 2001 before retiring in 2002. Three years later he died of prostate cancer, only two weeks before a meeting called "Archaea: The First Generation" that was being held in Munich in his and Stetter's honor. Claire Geslin, Évelyne's roommate on AMISTAD, turned out to be the most talented—and the most tenacious—in finally discovering the first virus infecting a *Pyrococcus*. This virus, called PAV1[18] (you can guess why), infects *Pyrococcus abyssi*, a strain that was isolated earlier in Prieur's lab. The infected strain produces particles in the form of a lemon (figure 2.5a), like SSV1, a virus that, as we saw in the first chapter, infects *Sulfolobus*. After the discovery of PAV1, Geslin, now an assistant professor in Brest, took on a graduate student, Aurore Gorlas. Gorlas was the first to successfully isolate a virus infecting a *Thermococcus*, *Thermococcus prieurii*, which she named in honor of her mentor, Daniel Prieur.[19] This virus, TPV1 (figures 2.5b, c), once again has the shape of a lemon. In the meantime, an Australian scientist, Mike Dyall-Smith, had discovered viruses in the shape of lemons that infect halophile archaea living in the great salt lakes of Australia. This viral form was thus present and very widespread among archaea. Recently, a young Lithuanian scientist, Mart Krupovic (who was also at our restaurant dinner), with the help of computer research on their proteins, showed that all these viruses

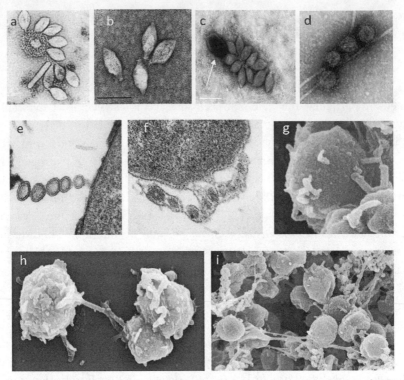

FIGURE 2.5. TEM of viruses and vesicles from *Thermococcales*. *A*, lemon-shaped virions of PAV1 attached to a vesicle. *B*, virions of TPV1. *C*, virions of TPV1 attached to a vesicle (*white arrow*). *D*, vesicles looking like virions of some eukaryotic viruses. TEM thin sections of cell producing chain of free vesicles (*E*) or vesicles surrounded by cell wall (nanotubes) (*F*). *G*, SEM of *Thermococcus* cell producing small nanotubes and of cells connected by long nanotubes (*H, I*). *A, B, C, D* courtesy of Claire Geslin and Aurore Gorlas (Université de Bretagne Occidentale, Brest, France); *D–I* courtesy of Évelyne Marguet (Université Paris-Saclay).

of archaea in the shape of lemons (or spindles) indeed belong to the same family: that of the *Fuselloviridae*.[20]

Gorlas has joined us at Orsay to work on the membrane vesicles produced by these microbes from hell. And yes, lacking viruses, all the strains we had brought back from the AMISTAD expedition produce vesicles. But let's go back ten years and meet Nicolas Soler, the first student to join us to work with Évelyne in her hunt for viruses. Nicolas is not here today, because he was offered a position as assistant professor at Nancy, in northeastern France. He was the first to begin working with Évelyne for his thesis on the samples from the AMISTAD expedition; his thesis subject was the search for viruses among *Thermococcales*. Évelyne explained that many of the viruses that infect archaea coexist with the infected cell (this is the carrier state). The archaea continues to divide while producing viral particles. The genome of the virus (always DNA in the case of archaea viruses) is present in the cytoplasm of the cell, where it is replicated at the same time as the cellular chromosome, but in a greater number of copies. The first step was for Nicolas and Évelyne to use classic techniques of molecular biology to identify among our strains of *Thermococcus* those possessing extrachromosomal DNA. The latter, generally smaller and more compact than chromosomal DNA, is obtained by causing the DNA of the strain being studied to migrate through electrophoresis onto a gel of agarose (a very purified agar) that enables macromolecules to separate according to their sizes and shapes.

A chromosomal DNA can correspond to a viral DNA, but it can also correspond to a plasmid, those little circular DNA found in many microbes. If the strains containing extrachromosomal DNA were indeed infected by a virus, they should produce viral particles. To separate these viral particles from the cells, the culture medium was centrifuged at very high speed: twenty thousand revolutions per minute (a superspinner). For this experiment, the culture medium containing cells and—we hope—viral particles is placed in large tubes of nitrocellulose, themselves placed in a rotor that turns inside the centrifuge. Subjected to gravity, the cells find themselves at the bottom of the tube, forming a pellet, whereas the viral particles, smaller and less dense, remain floating in the supernatant. After the viral particles are carefully removed (to avoid contamination with cells from the pellet), the supernatant is then centrifuged at an even higher speed (fifty thousand revolutions per minute) using an even more powerful centrifuge, an ultracentrifuge. In this case, gravity should be sufficient to cause the viral particles to fall to the bottom of the tube, mixed with small cellular debris. The pellet thus obtained is dissolved in a small volume of liquid and a drop of it is placed on a metal screen to be observed under an electron microscope.

When Cells Have Speech Bubbles

Nicolas and Évelyne thus spent hours and hours hunched over images provided by the microscope, scrutinizing the cultured floaters from the AMISTAD strains in search of those viral particles, without success. This work had some high points, however, because the virions indeed seemed to be there (figure 2.5d). In almost all the floaters they observed spherical structures surrounded by an envelope, greatly resembling some well-known viral particles, such as those of the SARS virus or the West Nile virus, but they weren't virions. The analysis of their composition in proteins showed that these particles were made up of proteins from the membranes of cells: they were membrane vesicles.

The sizes of these vesicles, ranging from 50 to 150 nanometers, corresponded to those of viral particles; some even contained DNA. Each time, hope was renewed: perhaps we finally had a virion and its viral DNA! Here, too, we were disappointed. It was only cellular DNA. Furthermore, Nicolas showed that the extrachromosomal DNA that we had detected in these strains did not correspond to genomes of viruses, but to plasmids. As time passed, we began to worry about Nicolas's thesis. He was supposed to isolate viruses, but we found only vesicles. What did those vesicles correspond to? Artifacts? Pieces of membrane from lysed cells that folded on themselves (a well-known property of lipidic membranes used by some scientists to manufacture artificial vesicles)? Diving into the literature, I realized that a small number of scientists had observed this phenomenon for a long time in bacteria. In particular, an American microbiologist, Terrance J. Beveridge, had spent a good part of his career studying them.[21] But his work had remained unknown to most bacteriologists and never found its place in microbiology textbooks. Many biologists were therefore skeptical of the importance of these vesicles. What if they were wrong? What if the massive production of vesicles that we observed day after day had a deeper significance? This production seemed to me to correspond to a true biological phenomenon and not only to an artifact of an experiment (figure 2.5e). Sometimes these vesicles were produced in clusters, where several vesicles head to tail were surrounded by a wall forming an elongated tube (figure 2.5f, g). We decided to publish our observations in 2007 in *Research in Microbiology*, a journal of the Institut Pasteur.[22] This article, associated with the description of several new plasmids of archaea,[23] would finally enable Nicolas to defend his thesis and our team to obtain funding from CNRS to pursue our research on vesicles.

In 2008 a new doctoral student, Marie Gaudin, who also dined with us at the restaurant, and an apprentice, Emilie Gauliard, came to join us and continue Nicolas's work. They would show, by taking thin slices from the cells of *Thermococcales* and observing them under an electron microscope, that vesicles form through budding in the cytoplasmic membrane. We then discovered in the literature that an analogous phenomenon had been observed in eukaryotic cells. These vesicles are called ectosomes or microparticles. The cells of the inhabitants of the three domains of life thus produce extracellular membrane vesicles.[24] In eukaryotes, as we also discovered in the literature, some vesicles are formed inside cells before being expelled to the outside— these are endosomes, and they closely resemble particles of the AIDS virus, and retroviruses in general. Some authors even wonder if retroviruses come from endosomes, or vice versa.[25] In 2010, the number of publications on vesicles began to explode as researchers discovered that cancer cells produce vesicles that are found in blood and can be used to diagnose a nascent tumor.[26] We wonder if the vesicles produced by cancer cells might play a role in the formation of metastases.

Some vesicles produced by human cells transport a genetic message in the form of small RNA that can regulate the activity of messenger RNA.[27] In fusing with another cell, this message can activate new functions. Membrane vesicles might then correspond to a new means of communication between cells. In multicelled organisms, such as humans, vesicles could enable communication among tissues. In bacteria, scientists have discovered that vesicles can transport antibiotics or other toxins to kill competitive bacteria. By contrast, the same vesicles can transport molecules that will serve as signals to organize life in a community of bacteria inside biofilms. A colleague compared vesicles to speech bubbles in comics and published a paper entitled "Microbiology: Bacteria Speech Bubbles."[28] In a bacterium from the ground, *Delftia*, American scientists observed the formation of clusters of vesicles very similar to those we observed in *Thermococcales*.[29] They call these structures "nanopods" (they are a nanometer thick). In two other bacteria, *Bacillus subtilis* and *Staphylococcus aureus*, Israeli scientists have observed that these nanopods (which they call nanotubes) can link several bacteria among them and transport proteins and perhaps plasmids from one to the other.[30] We also observed such nanopods connecting cells of *Thermococcus*, suggesting possible direct communication from cell to cell.[31]

Interest in vesicles has thus taken off these past few years. The International Society for Extracellular Vesicles (ISEV) was even created in 2010, with specialized journals and an annual meeting since 2011.[32] Unexpectedly, we find ourselves in the headlines with our vesicles. This certainly facilitated our task in obtaining funding. In fact, the European project EVOMOBIL that we are celebrating today at La Crémaillère focuses in part on the study of vesicles of archaea and of hyperthermophile bacteria, in particular on the role these vesicles might play in the transfer of genes in extremely hot environments.

We will see that the transfer of genes among archaea and bacteria has played an important role in the history of microbes from hell. It seems that it is the result of a transfer of this type that the reverse gyrase, which I mentioned briefly at the beginning of this chapter, was transferred from archaea to certain bacteria. We know that these transfers took place thanks to the comparative analysis of protein sequences. When a protein of a bacterium and a protein of an archaea have evolved from an ancestral protein that was present in the ancestor common to both domains, we observe many differences in the order of amino acids that form the sequences of these proteins.

That is not surprising, given that these proteins have diverged independently of each other for more than three billion years. We can say that there are two versions of the same protein, a "bacterial" version and an "archaeal" version. In general, all bacteria possess the bacterial version and all archaea possess the archaeal version. Yet sometimes a protein isolated from an archaeon corresponds to the bacterial version, which is a sign of a gene transfer; this protein of archaea has not evolved from a protein that was present in the ancestor common to archaea and bacteria, but it (or rather its gene) must have been directly transferred from a bacterium to an ancestor of the archaea in question (figure 2.6c).

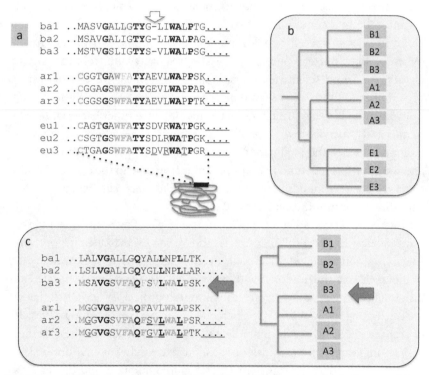

FIGURE 2.6. The three versions of universal proteins and lateral gene transfer. *A*, alignment of short portions of homologous universal protein sequences from three bacteria (B1), three archaea (A1), and three eukarya (E1). Each letter corresponds to an amino acid (aa) (e.g., A is alanine). Conserved essential aa that have not changed from the time of LUCA are in bold. They help construct the alignment (they should be in the same column). One aa is missing in bacterial sequences (*white arrow*), corresponding to a deletion in bacteria or to an insertion in a common ancestor of archaea and eukarya. The eukaryotic and archaeal versions are very similar (*common aa in bold gray*) (see chapter 5). *B*, a schematic tree obtained from the comparison of these sequences. *C*, alignment of short portions of homologous protein sequences from three bacteria (B1) and three archaea (A1). The sequence of the protein ba3 is much more similar (*bold gray*) to the archaeal sequences than to the other bacterial sequences and groups with archaeal sequences in the tree. This suggests that the gene encoding protein ba3 has been transferred from an archaeon to an ancestor of B3.

Analyzing the genomes of *Thermococcus*, an archaeon, and *Thermotoga*, a bacterium, using the method we've just described (comparing the sequences of their proteins) has revealed that hyperthermophile archaea and bacteria have exchanged many genes in the past.[33] Fragments of DNA can thus pass from one to the other in nature, but how? There are several well-known ways to transfer DNA between two organisms. These genes can be carried by viral particles that transported them in error; this is called transduction. But we saw in chapter 1 that the viruses that infect bacteria and archaea are very different from each other. They are also, in general, very specific to a small number of related species. Thus there is little chance that a viral particle containing bacterial DNA can inject that DNA into an archaeon, and vice versa. The most common mechanism to transfer DNA from one species to the other is transformation. When a bacterium dies and lyses, it releases DNA into the environment in the form of chromosome fragments or small circular DNA: the famous plasmids. These DNA can be recovered by another bacterium if it possesses a DNA pump. This is the way that genes resistant to antibiotics can be transmitted among bacteria in hospitals. But DNA freed in nature at 80°C or more will not survive very long. We will see in the next chapter that such DNA will be denatured (the two strands of the double helix will separate) and degrade (the long DNA molecule will break into small pieces). How can we imagine *Thermococcus* and *Thermotoga* being able to exchange genes in their burning environment? This is where our extracellular membrane vesicles come into play.

In his thesis, Nicolas Soler compared the fate of free cellular DNA and DNA associated with membrane vesicles produced by *Thermococcales* at 90°C. The results were very clear: whereas free DNA was completely degraded after being exposed for 30 minutes to 90°C, DNA associated with vesicles remained practically intact.[34] Is it possible that vesicles serve as vehicles for the transfer of genes among microbes from hell in hot springs? This is still just a hypothesis, but in her thesis Marie Gaudin has shown that in the laboratory, vesicles are able to transfer genes between two cells of the same species, *Thermococcus kodakaraensis*.[35] This strain, isolated by Japanese colleagues on the island of Kodakara, is capable of recovering DNA through transformation (it must possess a DNA pump). Thus Marie was able to introduce a plasmid containing a genetic marker into *T. kodakaraensis*; she observed that this strain now produced vesicles containing plasmid. She then purified these "plasmidic" vesicles and mixed them with a strain of *T. kodakaraensis* not containing the plasmid. What happened? Thanks to the genetic marker that enabled it to be traced, Marie was able to observe that some strains of *T. kodakaraensis* had recovered the plasmid in question, no doubt transported by vesicles.

Do such events occur in nature, at the bottom of oceans or in the heart of Icelandic mountains? We don't know yet. But Nicolas and Marie have made other very interesting observations. They have shown that vesicles could naturally contain plasmids. This is the case in particular with our favorite strain, *Thermococcus nautili*, which was baptized thus in memory of the AMISTAD expedition and Évelyne's dive with the *Nautile*. This strain contains three plasmids, called pTN1, 2, and 3 (here, too, you can guess why). When Marie isolated the vesicles produced by *T. nautili*, she saw that some of them contained plasmid pTN3.[36] When we obtained the DNA sequence of pTN3, we noticed that it was the genome of a virus. Indeed, this sequence included a coding gene for a protein typical of a virus, a capsid protein, the shell that forms the structure of the viral particle. This protein strangely resembles the capsid protein of Adenovirus, a virus that infects humans and provokes varied illnesses, from pneumonia to gastroenteritis to conjunctivitis. The same protein is found in a virus called PRD1, which infects the bacteria *Escherichia coli*. We will return in chapter 4 to these resemblances that bear witness to a kinship link among viruses infecting organisms belonging to different domains and that testify to the antiquity of the virus.

What was remarkable in the case of the membrane vesicles transporting pTN3, a plasmid bearing the gene of a viral capsid protein, is that we don't find this protein in the vesicles in question. Marie has shown that their envelope is formed by membrane proteins coded by the genome of the cell. These vesicles transporting pTN3 are thus not virions because, by definition, virions are formed by one or several proteins coded by the genome of the virus, itself present in the virion. We have proposed the term "viral vesicle" for this new type of biological entity. Why is this observation important? Is it limited to viruses from hell? It seems not. At the beginning of 2014, an American team from the Massachusetts Institute of Technology (MIT) published a paper in *Science* describing the abundant presence of bacterial membrane vesicles in the ocean. These vesicles contain DNA that, according to the authors of the paper, corresponds essentially to bacterial DNA. Still, continuing the analysis of their data with Mart Krupovic, we observed that a good part of this DNA probably came from plasmids or viruses infecting bacteria.[37] The viral vesicles would thus be an essential component of the environment.

And in humans? We have seen that our cells also produce vesicles. Scientists have shown that human cells infected by a herpes-type virus produce vesicles that contain proteins and viral messenger RNA,[38] so why not viral genomes? This remains a hypothesis, but we can imagine that cells infected by a retrovirus, like the AIDS virus, might produce vesicles containing the genomic RNA of the virus, which would enable it to be transmitted, sheltered from the immune system of the host. This explains, among other things, why we still don't have a vaccine. These are perhaps only ramblings, because I am not a specialist in this area, but the concept of viral vesicles is pleasing . . . for the moment.

In my opinion, the repercussions of Évelyne's voyage in the *Nautile* will not end soon. Future pages will be written by newcomers recruited thanks to the ERC funding, such as Sukhvinder Gill, Ryan Cathpole, and Morgan Gaia. Before joining us, Morgan worked in Marseille on his thesis on small viruses that infect giant viruses (see below). Didier Raoult has christened these unusual viruses of viruses "virophages"[39] (formerly called satellites).

Looking around me in La Crémaillère, I see nearly thirty of us around the table sharing good food. A few scientists have brought their partners. Among all these people, one is particularly dear to me: David Prangishvili, whose bald pate contrasts with my head of thick white hair. How did he find himself seated at this restaurant in a suburb of Paris? That is another story I am happy to tell. But for that we must go back about ten years to another dining table, this time at a family home.

2003: At the Home of David and Nana in Regensburg

When we were beginning to worry that we wouldn't be able to isolate viruses of *Thermococcales* at Orsay, Évelyne and I said to ourselves, let's pay a visit to the king of viruses of archaea, David Prangishvili, to learn a few tricks that might help us in our quest for viruses. David is one of the pioneers of research on archaea; he began in the Soviet Union around the same time I did in Paris. We were even competitors for a time, each seeking to isolate the DNA polymerase of archaea on either side of the iron curtain. After the fall of the Soviet Union in 1992, when civil war erupted in Georgia, research in that country collapsed. David emigrated to Germany to work with Zillig in Munich, until Zillig retired in 2000.[40] David was then taken in by Karl Stetter in Regensburg. In Munich David became enthralled by Zillig's viruses, and finally picked up the torch of research on viruses from hell after settling in Regensburg.[41] In a set of successful expeditions, David and his students travelled from Italy to Yellowstone (figure 2.7a), discovering amazing new viruses from hell producing never-before-seen types of virions. One of them invented the clamp well before carbs did it (figure 2.7b); another produces bottle-shaped virions, corresponding to a new viral family, *Ampullaviridae* (figure 2.7c).[42]

We met David and his wife, Nana (the name of one of the great long-ago

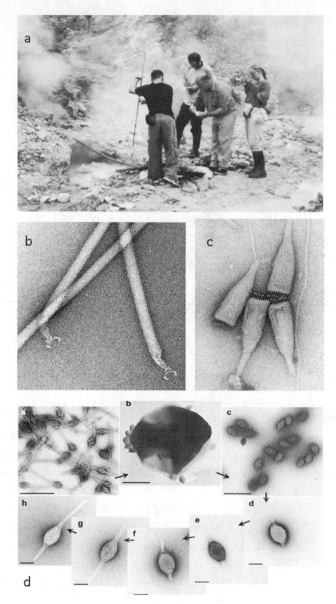

FIGURE 2.7. The wonderful world of archaeal viruses. *A*, David Prangishvili (*center*) with students sampling a Pozzuoli hot spring (Italy), courtesy of David Prangishvili. *B*, TEM of virions with claws produced by virus AFV1. The inset shows AFV1 virions attached to cellular pili via their claws, courtesy of Soizick Lucas, Institut Pasteur. *C*, TEM of bottle-shaped virions produced by ABV, discovered in Pozzuoli (modified from Häring et al., 2005b). *D*, virions of *Acidianus* two-tailed virus (ATV) at different stages of development (modified from Häring et al., 2005a). Lower TEM picture, extracellular modification of ATV virions: *a*, two-tail virions, *b*, extrusion of tailless virions from an ATV-infected *Acidianus* cell, *c*, tailless virions, *d–h*, purified virions incubated at 75°C for 0, 2, 5, 6, and 7 days. Scale bars: a-c, 0.5 microns; d, 0.1 micron.

queens of Georgia), during our stay with Zillig in Munich in 1999. We shared many steins of beer at the Oktoberfest in Munich, in a highly convivial atmosphere. So we returned to see David in 2003, this time in Regensburg. He had just made a spectacular discovery with Monica Hering, his German student at the time, which earned her the honor of publication in the journal *Nature*.[43] They had discovered a virus whose virions change shape after leaving the cell, which had never been observed before (figure 2.7d). At the beginning, this virus (called ATV, for *Acidianus* Tailed Virus) vaguely resembles a lemon, but later, if it is incubated at a high temperature, two tails appear on each side of the viral particle. After spending the day in David's lab, we went to his home, where Nana had prepared a wonderful Georgian meal accompanied by a bottle of red wine from Kakheti (in eastern Georgia).

After reminiscing about the best days of our previous encounters, I got to what was bothering me at the time. I had just been invited to join the Institut Pasteur in Paris to create a new unit and become head of the department of microbiology. A gift, you might say—an opportunity not to be missed. For someone who had always fought for the recognition of his favorite microbes, what could be more exciting than to have archaea enter the most prestigious of French institutes of research in biology? Yes, but Évelyne and I had just completed the installation of our platform for culturing anaerobic hyperthermophiles in Orsay. Must we abandon it? For Évelyne, it was out of the question. To transfer it to the Institut Pasteur seemed complicated, given the tight security at an institute in the heart of Paris. (Don't forget that our microbes from hell produced hydrogen sulfur.) In between the khinkali, a khachapuri, and a satsivi, three delicious Georgian specialties, I had an inspiration. Why not keep the Orsay lab and open a new one at the Institut Pasteur? Of course, one couldn't direct two laboratories in two different institutes, especially if one hasn't won a Nobel Prize, but I could still try to get past the bureaucratic hurdles. There remained the problem of feasibility: how could I direct two laboratories without being cloned? This is where David came in. Would he be interested in coming to work with me at Pasteur to pursue his work on the viruses of archaea? He knew of my love for these viruses and the reputation of the Institut Pasteur, and said yes right away. To my great surprise, Philippe Kourilsky, director of the Institut Pasteur at the time, immediately accepted my proposal, the Institut hired David, I kept my position as professor at Orsay (and the lab that went with it), and I started a new unit at Pasteur, with David as my right hand. I seized the opportunity to have Soizick Lucas, who had worked with us at Orsay on perfecting genetic tools for *Thermococcales*, and Simonetta Gribaldo, an Italian postdoctoral fellow who shared my passion for evolution, join us at Pasteur.

The Wonderful World of Archaeal Viruses

Once at Pasteur, I divided my large director's office into two to share it with David. I was delighted to be on the front lines to assist in the discovery of new viruses. I was very proud to enable the line of research that Zillig had initiated in the 1980s in Munich to be continued at the prestigious Institut Pasteur. I wouldn't be disappointed. The very favorable working conditions at the Institut, located in the heart of Paris, would enable us to invite several talented young scientists who would cut their teeth on viruses from hell. Two doctoral students would discover a new phenomenon in the world of viruses: the construction of pyramids. A student from the École Polytechnique, Ariane Bize, was the first to see these famous pyramids in thin section electron micrographs of *Sulfolobus* cells infected by the rudivirus SIRV2.[44] These seven-sided pyramids are formed on the surface of infected cells and pierce their envelopes (figure 2.8a).

Each pyramid then opens like a flower to form a hole on the surface of the cell, a hole through which the virus will expel its virions (figure 2.8b, c). This process had never been observed for viruses infecting bacteria or eukaryote cells. We were again facing a phenomenon specific to the third domain of living beings on Earth: their viruses. After Ariane, a Dutch student, Tessa Quax, was able to identify, then purify, the viral protein responsible for the formation of these pyramids. It involves a small single protein that is capable of self-assembly on the internal surface of cellular membranes to form these stupefying seven-sided pyramids.[45]

Even more impressive, Tessa succeeded in expressing this protein in a bacterial cell (a genetic manipulation where one introduces the viral gene into a bacterium to force it to produce the protein coded by the gene). One then sees pyramids form on the surface of the membranes of bacteria; unlike what happens with *Sulfolobus*, they are incapable of piercing the bacterial envelope, probably due to peptidoglycan (figure 2.8d). We saw in the previous chapter that peptidoglycan (the target of penicillin) is specific to bacteria. Would peptidoglycan, which one often interprets as a bacterial exoskeleton, be above all a protection against viruses? Does its presence in almost all bacteria explain why the diversity of viruses infecting bacteria is much smaller than that of viruses infecting archaea? This is what we, along with David, thought. More generally, we think that the arms race between viruses and cells is one of the major keys in biological evolution.[46] We know that wars have often brought decisive progress because when it is a matter of life and death, bureaucratic and financial hurdles that limit the creativity of scientists quickly disappear. Nuclear energy, reaction engines, and radar are the fallout of the Second

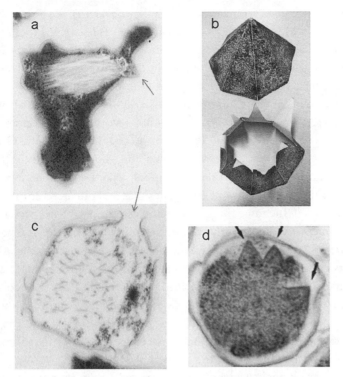

FIGURE 2.8. *A*, TEM thin section of SIRV (*Sulfolobus islandicus* rod-shaped virus) infected *Sulfolobus* cell (virocell), producing pyramids that cross the cell envelope (*gray arrow*). Numerous virions (long rods) accumulate in the cytoplasm. *B*, paper models of closed and open seven-faced pyramids. *C*, TEM thin sections of an empty SIRV virocell with an open aperture after pyramid opening (*arrow*). *D*, a cell of a recombinant bacterium *Escherichia coli* producing pyramids from the archaeal virus SIRV2. They cannot pierce the peptidoglycan-containing cell wall and remain in the periplasm. (*A, C, D*) Courtesy of Ariane Bize and David Prangishvili, Institut Pasteur.

World War. Humankind's landing on the moon would probably not have happened without the Cold War. It's sad but true. The same thing happens in the living world: the struggle between viruses and cells is often a struggle to the death, even if we know of many examples of the two coexisting. In any event, it is a struggle that everyone takes seriously.

The viral protein capable of constructing a pyramid teaches us something important: we mustn't underestimate viruses. They are capable of great things—in particular, of inventing new proteins and the corollary, new functions. Notably, a very interesting property of viruses is their ability to integrate their genome into that of their hosts. Sometimes these integrated genomes give birth to new viruses, which is not always very pleasant. But most often, they will degrade and disappear, unless the infected cell finds a benefit in using a viral protein coded by these genomes on the path to disappearance.

This is what happened for mammals when they began to use a gene of a retro-virus (like AIDS) to manufacture the placenta.[47] This enabled them to keep the embryo inside the mother's body (there was no longer a need to lay eggs) because this gene codes for a viral protein that is fixed on the surface of the placenta and prevents its rejection by the immune system.

The genomes of all cellular organisms are filled with intact or more or less degraded viral genomes. In the Institut Pasteur's BMGE team, Diego Cortez, a Mexican doctoral student who worked with Simonetta Gribaldo, perfected a bioinformatical technique enabling the detection of integrated viruses or plasmids in bacterial and archaeal genomes. Thus he was able to reveal an uninterrupted flow of viral genes that accumulate over time in cellular genomes.[48] The same thing is true of plasmids, which are probably former genomes of viruses that have lost the ability to produce viral particles. Viruses and plasmids form what we sometimes call mobile elements. All these elements gravitate around cellular lineages that they parasite; this is called a mobilome. The preliminary work Diego carried out at Pasteur in collaboration with Nicolas at Orsay on the mobilome of *Thermococcales* has shown that the latter coevolves with archaea. This is why in the EVOMOBIL project we want to retrace the history of the mobilome of archaea on a large scale and see what role vesicles, viral plasmids, or others play in this story.

I have already spoken a lot about viruses, but I have not raised a question that often comes up in discussions about them: Are viruses living? Are they really microbes or are they simply chemical machines that use biological information to reproduce? For a long time, most biologists favored the second hypothesis, because they associated viruses and virions. The question was raised again following the 2003 discovery by Didier Raoult's team in Marseille of a giant virus, the Mimivirus, whose virions are so big they can be seen under an optical microscope.[49] These giant viruses, which infect amoeba, eukaryotic cells that are themselves giant, were for some time confused with bacteria. The sequencing of their genomes has shown, however, that they do not code for ribosomes, so they are indeed viruses. How can we deny such viruses the status of living organism?[50] Also, we have just seen that viruses are a major source of genetic novelties. New genes appear constantly in viral genes during evolution. Viruses have thus "invented" many proteins, such as the one we have just been talking about, capable of manufacturing pyramids. Finally, viruses are very abundant; they infect all cells. Molecular ecologists (yes, they exist) have shown that the greatest part of the genetic information we find in the environment is of viral origin. It seems difficult to accept that nonliving molecular machines produced all this genetic diversity.

We can resolve this problem if we realize that a virus expresses its "living" character only once its genetic material is present in the infected cell. The cell is transformed into a "viral cell"—or virocell—by the information the virus brings.[51] To paraphrase the metaphor of the French Nobel laureate François Jacob, whereas the dream of a prokaryotic or eukaryotic cell is to produce two cells, the dream of a virocell is to produce as many virions as possible to transfer its genetic information to the greatest possible number of cells.

When the SIRV2 virus infects a cell of *Sulfolobus*, it degrades its chromosome and replaces it with its own genome. The membrane is covered with pyramids; the virions accumulate in the cell while waiting for the pyramids to open. We are still dealing with a cell, but it is no longer an archaeon, since it no longer contains the genes of an archaeon; it is a virocell. At least that's my opinion. It is not shared by all my colleagues, and the idea is sometimes the subject of impassioned debates.

We will talk about viruses again when we discuss the place of microbes from hell on the universal tree of life. We will ask whether viruses have their place on this tree, which will lead us to again ask the question of their profound nature and their origin. Before we do that, it is time to look at our microbes from hell and their viruses with the tools of biochemistry to try to understand how these extreme organisms manage to frolic in their baths of boiling water. The next chapter will thus be a bit technical, but we must soldier on if we truly want to understand our new microbial friends.

How Do You Live in Hell?

Can there be life at high temperatures? For a long time, scientists were skeptical of this possibility. Why? In chapter 1 we discussed the fragility of proteins. But other molecules or structures essential to life are also clearly unable to tolerate infernal temperatures: normally, DNA is unable to function at a temperature higher than 70°C, and the membranes that control the flow of molecules entering or leaving cells normally become true sieves above 80°C. How have hyperthermophiles been able to overcome all these obstacles? To answer this question, let's take a little detour through chemistry.

No Life without Liquid Water

When we ask what conditions are necessary for life, it is generally the presence of water that immediately comes to mind. This magical molecule is abundantly present in all our cells, and it is water that astrobiologists seek first when they attempt to identify planets where other forms of life might have appeared. There is no life without water for several reasons, all of which are connected to the physicochemical properties of that molecule. One of those properties will be the focus of our attention here: water possesses the remarkable ability to join forces in a reversible and "plastic" way with other molecules in a living organism and to facilitate interactions among those molecules. These interactions are essential to the formation of the "weak bonds" that we will soon discuss in some details. These bonds are easily and rapidly reversible, which is critical for interactions between molecules in living organisms. However, these interactions are possible only if the water is in a liquid state. Caught in ice, the molecules of an organism would be incapable of moving and interacting. On the opposite end of the spectrum, in vapor, all water molecules would move away from each other and go off in many directions: it is difficult to envision encounters under those conditions. And so, is it the temperature at which water boils that determines the temperature above which life can no longer exist on our planet?

It is not that simple. When pressure increases, water remains liquid above 100°C. However, we have seen that the upper limit for life on Earth is around 110°C. To understand what prevents life from going beyond that limit, let's take a look at different types of molecules that make up a living organism and the mechanisms that enable them to interact.

Carbon: The Key to Life

Of the hundred or so atoms chemists have identified in the universe, only about twenty are found in the living world, and only five of these are present in great abundance and form almost all the molecules specific to living beings. These are carbon (C), oxygen (O), hydrogen (H), nitrogen (N), and, to a lesser degree, phosphorus (P).

What a contrast with the incredible diversity of living things! The variety of organisms that swarm over the earth, under the sea, and in the air seems almost infinite. They couldn't all have fit into Noah's ark. We can already count probably several million species of insects, and those make up only a small part of the animal kingdom, which itself is a very small subset of the living world. Although we don't always realize it, the diversity of species of single-celled microbes, archaea, or bacteria is much greater than that of all the multicelled organisms (animal and plant, in particular) combined. Furthermore, within a single species, the variety of individuals is again astonishing. Just think of all our human brothers and sisters: not one of the seven billion inhabitants of our planet is identical to another. All this diversity is associated with a complexity that one finds only in living beings: the smallest virus is already much more complex than any mineral. This is truly amazing.

To this incredible visible diversity in the living world there corresponds a diversity on another scale: that of proteins and nucleic acids, DNA and RNA. With the five principal atoms listed above, living beings are able to manufacture an almost infinite number of different molecules, some of which are also of great complexity. How is this possible? It is due to the unique properties of the carbon atom. We might even say that this atom, by its very nature, has enabled the appearance of life on our planet. Why? After all, when it is in a pure state, carbon forms the diamond (if it has been compressed in the depths of Earth) or graphite (coal, in some sense): there is nothing very alive in that. It is only when it is in the company of the four other "atoms of life" (O, H, N, and P) that the carbon atom reveals its magical properties. Here we have touched on the essence of the mystery that distinguishes living matter from inanimate matter: only the carbon atom enables the formation of molecules that can be simultaneously extraordinarily varied and complex. It can join forces with the four other atoms through strong bonds (chemists would say that the valence of carbon is four). For example, in the methane molecule, whose chemical formula is CH_4, the carbon atom is bonded to four hydrogen atoms.

Chemists call the bonds that link atoms in a living being's molecules "covalent." They are very solid, which enables them to combine with many atoms, and thus to obtain molecules that can be very long and have a high potential for interaction. For example, a great number of carbon atoms can assemble linearly to form a long chain. In this case, each carbon atom retains the possibility of being bonded again with two other carbon atoms, nitrogen, oxygen, or hydrogen. All those atoms have a valence inferior to that of carbon (3, 2, and 1, respectively). But when linking these four atoms, following the rules of valence, there can be an infinite number of carbon-based organic molecules. We generally draw chemical molecules in a two-dimensional way, but in reality they of course exist three-dimensionally. Thus the four atoms that can combine with carbon are found on the four points of a pyramid of which the carbon atom occupies the center. This enables the construction of an infinite number of very large molecules, called macromolecules, in three-dimensional space. To discover the secrets of hyperthermophiles, let's observe the effects of infernal temperatures on these giant molecules that are characteristic of the living.

Covalent Bonds: Solid Bonds . . . Up to a Point

While observing the fate of the contents of an egg when it comes in contact with a burning pan, I noted in chapter 1 that the macromolecules that make up the egg don't tolerate heat very well. Let's look at this a bit more closely. Macromolecules are formed by the linear assemblage of a large number of small organic molecules, called monomers. There are a limited number of monomers: twenty amino acids for proteins, and four nucleotides for nucleic acids (DNA or RNA). But thanks to the chemical properties of carbon, these monomers can bond together in a regular formation, one behind the other, joined by covalent bonds to form polymers that can contain hundreds, thousands, even millions of atoms, in the case of DNA molecules. By altering the number or nature (or both) of the monomers, one can obtain an infinite number of combinations, exactly as we are able to write an infinite number of different novels with the twenty-six letters of the alphabet. The number of different proteins and nucleic acids that exist or have existed is almost infinite and, like the Greek god Proteus for whom they were named, proteins can assume every possible and imaginable shape to fulfill their role in the great game of life.

At the foundation of all macromolecules we always find covalent bonds. In such a chemical bond, electrons—which "rotate"[1] independently around several nuclei when two atoms are isolated—come together to form a single electron cloud that surrounds the nuclei of these two atoms when they are linked by a covalent bond (so the two atoms put their electrons "in common"). This is why covalent bonds are very strong. To create them (to force the electrons to cohabitate) takes energy. But once they are formed, they are generally very stable at "low" temperatures. If placed at below 100°C, the rupture of the covalent bonds almost always necessitates the action of specific proteins: enzymes. Enzymes are biological catalysts: each one considerably accelerates the speed of one of the many chemical reactions that delimit the life of an organism in very precise temperature or pH conditions.

By contrast, if the temperature exceeds a certain threshold, all the strong bonds that ensure the cohesion of organisms ultimately break. This is what happens, for example, when living matter burns: after a certain amount of time it is carbonized, reduced to a pile of carbon atoms. No life is possible at temperatures at which covalent bonds are broken. This explains in particular the absence of microbes living at 250°C. Let's recall the experiment that put an end to the controversy (see chapter 1): the American biochemist Robert White had shown that at that temperature, proteins were degraded irreversibly within a few minutes.[2] In other words, the covalent bonds between amino acids were broken one after the other. At too high a temperature, we observe a degradation of macromolecules through a rupture of strong bonds: this is called thermal degradation. Thus a major constraint is imposed on living beings: they can only explore and colonize environments in which the covalent bonds formed by carbon and its partners are stable.

For this reason, we estimate that no life is possible at temperatures above 150°C. In this case, you might say, why don't we find any superhyperthermophiles capable of growing between 100 and 150°C? There are several reasons for this. First, as I've already suggested, the "strong" bonds are not always as strong as you might think. We will discover that some organic molecules possess a true Achilles' heel at high temperatures: a relatively fragile covalent bond that risks breaking before 150°C, even sometimes well below 110°C. Furthermore, covalent bonds are not the only ones that contribute to the construction of the living being—far from it. Alongside them, other types of so-called weak chemical bonds even play principal roles to explain the extraordinary alchemy of the living.

Not Too Strong, Not Too Weak: The Subtle Alchemy of Weak Bonds

Let's look again at the fried egg experiment in chapter 1. At the moment when the white and the yoke fell onto the burning bottom of the pan, the proteins—we have compared them to an open pearl necklace rolled like a ball of string; now we know those pearls are monomers, amino acids—unfold through the agitation of the atoms of the metal that forms the bottom of our pan. This agitation, which was provoked through heating, is transmitted to the atoms that make up the proteins of the egg. The reaction is instantaneous and is produced at a temperature well below 150°C. In fact, the bonds that maintain the folded necklace of amino acids are not covalent (strong) bonds, but weak bonds (see box 3.1). In other words, if strong bonds create the skeleton of a protein, its folding in space is owed to the formation of weak bonds between amino acids that appear at a distance from each other when we focus on their linear chaining in the polypeptide chain. And so, the shape of a protein depends on the number and nature of these bonds, and thus on the chaining order of the amino acids—the "sequence" of the protein in the jargon of biochemists—along a chain that can include several dozen or sometimes even several hundred amino acids. Note that weak bonds are precisely those that reversibly connect water molecules in liquid water and ice (figure 3.1).

The difference between these two types of bonds is essential. In a weak bond, we do not observe a share of electrons, but "simple" attraction by the electron clouds of two atoms (or groups of atoms). This attraction is possible only if these two clouds do not have the same electrical charge (an attraction between the plus and the minus, in a sense; see box 3.1 for more information).

Beyond proteins, the spatial structure of DNA and RNA is also based on weak bonds—as are, as we will see, the biological functions of these macromolecules. It is weak bonds that enable biomolecules (large or small) to meet, touch, and separate, to find each other again (for example, during the encounter of an enzyme and its substrate, or of an antigen and its antibody). On a larger scale, weak bonds are again at the origin of macromolecular joining and the formation of very stable giant molecular edifices, such as ribosomes or cellular membranes or viral particles. These assemblages ultimately enable the formation of cells, tissue, and an entire organism. Thus it is no exaggeration to assert that weak bonds, in a dynamic balancing act, hold together all parts of the whole of each individual (microbe or elephant). These bonds are almost always much more sensitive to the thermal agitation of atoms than are strong bonds: they don't like heat. As a result, at high temperatures macromolecules, whose shape in space is assured by weak bonds, become denatured, just like proteins (this is called thermal denaturation).

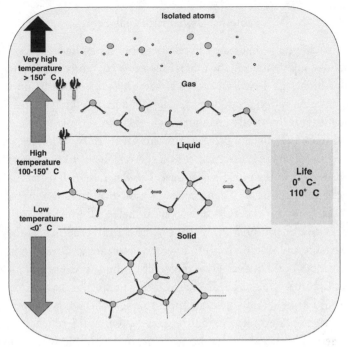

FIGURE 3.1 Effect of temperature on molecular organization. Schematic molecules of water (H_2O) made of two atoms of hydrogen (*small circles*) and one atom of oxygen (large circles) linked by strong covalent bonds (*thick lines*). At temperatures compatible with life, they also form transient weak hydrogen bonds, producing a liquid state. If temperature increases, these weak links are first broken and a gas is formed (vapor). If temperature increases further, covalent bonds themselves can be broken. But if temperature decreases, all molecules are linked by noncovalent bonds that become stronger, forming a crystalline-like network (ice).

Here we have put our finger on one of the principal problems encountered by hyperthermophiles: they live at temperatures at which weak bonds are . . . too weak. More generally, it is this type of bond that in large part determines the range of temperatures compatible with life.

If it is too cold, the weak bonds become too strong: the movement of atoms is no longer capable of breaking them, the interactions between macromolecules become irreversible, and the cell ends up "freezing," like water below 0°C. If it is too hot, the weak bonds are no longer sufficient to maintain the shape of the macromolecules and ensure the interactions among them; the entire cellular edifice risks collapsing (not to mention the risks of thermal degradation mentioned above). Temperature is clearly a crucial parameter in the equation of life. In our quest for the secrets of hyperthermophiles, we will now observe under the microscope the behavior of chemical bonds (strong and weak) in proteins, DNA, and RNA when temperatures become infernal.

Proteins: Giants with Clay Feet

We saw in chapter 1 that many hyperthermophile archaea do very well at temperatures above 100°C. The official record is currently held by *Pyrolobus fumarii*, which still divides at 113°C. However, well below this upper limit, normal proteins suffer: they "unfold," they are denatured (again, think of the experiment with the fried egg; figure 1.4). This is a serious problem for our microbes from hell, for if a protein "unfolds," it is really no longer a protein. This question of shape is essential: it is at the very heart of the most intimate mechanisms of a living organism. Indeed, the way in which a chain of amino acids folds on itself in space determines its biological function. In particular, it leads to a grouping of certain amino acids that, on the surface of a protein, creates cavities or protuberances that enable it to interact with other molecules or macromolecules. Thus for an enzyme to catalyze the formation of a strong covalent bond between two molecules, it must possess on its surface a cavity, called an active site, capable of receiving them (the classic image is that of a lock and its key) and accelerate the formation of the bond in question.

Temperature does not affect in the same way the different types of weak bonds (hydrogen bonds, ionic bonds, and hydrophobic interactions, see box 3.1) that ensure the shape and thus the function of proteins. Among mesophilic organisms (which live at "moderate" temperatures), above a maximum temperature that can vary between 50 and 70°C depending on the molecule and the organism, the thermal agitation of the atoms gradually breaks the hydrogen bonds, then the ionic bonds. By contrast, hydrophobic interactions are reinforced at high temperatures, but overall, this cannot compensate for the loss of hydrogen and ionic bonds. The chains of amino acids thus unfold: The proteins become completely inactive; they are denatured. And their plight has not ended. In these conditions, the residues of hydrophobic amino acids, normally hidden within proteins, are exposed to water molecules (which they don't really like). To protect themselves from water, hydrophobic residues of an unfolded chain will then interact with those belonging to other unfolded chains to form a solid, three-dimensional network: the proteins will then coagulate (this is the principle of the omelet or fried egg).

How have hyperthermophile microbes managed to overcome such obstacles?

Proteins That Do Not Coagulate in Hell

We have seen that in 1903, the American scientist William Setchell asked why the "protoplasm" of the inhabitants of the hot springs in Yellowstone didn't

BOX 3.1. WEAK BONDS: ANOTHER GIFT FROM THE
ELECTRICITY FAIRY

Weak bonds between amino acids are formed on the level of chemical groups—which biochemists call residues—characteristic of each of the twenty amino acids. Not being involved in the formation of the covalent bonds between amino acids (peptide bonds), these residues are free to participate in other interactions among the monomers of proteins (figure 1.4). Three large types of weak bonds can thus be formed, depending on the nature of the chemical groups involved:

- Ionic bonds (the "strongest," or the least weak) associate amino acids whose residues possess acidic or basic groups carrying opposing electrical charges (negative for the acid residues, positive for the basic residues).
- Hydrogen bonds involve amino acids whose residues possess an alcohol functional group OH or SH. The oxygen atom (O) and the sulfur atom (S) of the latter carry a negative electrical charge (called partial, because the SH and OH groups are neutral overall) and can form electrostatic interactions with a partial charge of positive sign carried by the hydrogen atom (H) of another amino acid. Recall that hydrogen bonds become more intense as the temperature lowers. Thus it is hydrogen bonds between water molecules that explain why these molecules remain loosely associated in liquid water. However, at low temperatures (below 0°C) the hydrogen bonds become sufficiently stable to immobilize water molecules, forming ice crystals (figure 3.1).
- "Very weak" bonds are formed between residues rich in carbon and hydrogen. They are also electrostatic and are called van der Waals type, named after a Dutch chemist at the end of the nineteenth century who discovered them. Interactions of the van der Waals type are often called hydrophobic ("that which doesn't like water") because the residues involved are incapable of bonding with water molecules.

By contrast, the residues that are electrically charged can establish bonds with these H_2O molecules that carry a partial negative charge on the oxygen atom and a partial positive charge on the level of hydrogen atoms. These are called hydrophilic ("that which likes water") residues. The most stable folding of the chain of amino acids of a protein generally corresponds to a more or less globular shape, where amino acids with a hydrophilic residue are oriented toward the exterior (toward water molecules) and the amino acids with hydrophobic residue are oriented toward the interior (giving the impression they are fleeing water).

coagulate. Indeed, if we plunge a hyperthermophile microbe into a cup of boiling water, it remains unchanged (unlike what we observe with an "ordinary" microbe). Its proteins have not been denatured. Some of them can be heated to 120 or even 140°C without losing their shape or their activity.

Well-Sheltered in an Electrostatic Cage

So do microbes from hell possess a new type of protein? Amino acids capable of establishing novel bonds? Not at all. Their proteins are formed by the same amino acids as ours, and there doesn't exist in nature a chemical bond unique to hyperthermophiles. Are their proteins folded with the help of strong bonds? This isn't the case either. If we really think about it, the stabilization of proteins by strong bonds would be a dead end: to create such a bond demands a lot of energy and requires the activity of a specific enzyme (thus of a protein). To create a particular protein, dozens of enzyme proteins would be necessary, each one specific to one of the strong bonds that would give it its shape. But things wouldn't stop there, since to create each of these protein enzymes, dozens of protein enzymes would again be necessary, and so forth.

After a great deal of investigation, the enigma of hyperthermophiles' resistance to high temperatures was finally resolved through a detailed study of their proteins through X-ray diffraction. The method consists of bombarding the molecule with X-rays and observing how it "bounces" on its various atoms. Although one needs to be strong in mathematics and physics to understand the principle of X-ray diffraction, one only needs good eyes to appreciate the results. X-ray diffraction indeed enables one to determine the position in space of each atom of a protein. One can then identify all the weak bonds between residues of amino acids and thus determine the spatial structure of the polypeptide chain.

One of the first analyses using X-ray diffraction conducted on a protein extracted from a hyperthermophile microbe was carried out in 1996 by David Rice, an English scientist at the University of Sheffield. I attended the lecture during which he presented this work at a meeting of European scientists entitled Biotechnology of Extremophiles. Rice compared the structure of the same metabolic enzyme (glutamate dehydrogenase) in both hyperthermophile (*Pyrococcus furiosus*) and mesophilic microbes.[3] The results he obtained were very clear, and the solution to the enigma of infernal temperatures "discovered" to be incredibly simple. The *Pyrococcus* protein had the same shape as its mesophilic sibling: the chains of amino acids were folded in practically the same way in the two microbes. To see the difference, one had to observe the enzymes more closely. One could then see that the surface of the *Pyrococ-*

cus protein contained many more amino acids with acidic and basic residues, carriers of negative and positive electrical charges. Thus networks of ionic bonds enclosing the enzyme like an invisible net were formed. On the projection screen, the difference between the protein of *Pyrococcus* and the "classic" protein was visible to the naked eye. David Rice had, in fact, following the tradition of biochemists, colored the acidic residues in red and the basic residues in blue. The hyperthermophile protein was much more striking, with the red and blue spots forming a "patchwork" on its surface.

Other labs then conducted other structural studies of pairs of hyperthermophile and mesophilic proteins and confirmed the observations made by Rice and his colleagues. Most of the proteins of hyperthermophiles that were analyzed presented an excess of charged amino acids and the ionic networks that go with them. Also, they observed in these proteins that subtle modifications in structure reinforce the stability of "proteins from hell" at high temperatures: here, a few supplemental hydrogen bonds; there, a hydrogen bond replaced by a hydrophobic interaction more stable at high temperatures; elsewhere, a region whose structure is particularly fragile in the mesophilic protein is completely eliminated; and so forth. Often we also observe that the hyperthermophilic protein adopts a more compact shape, in which its hydrophobic core is reinforced. In short, the proteins of microbes from hell are simply stabilized by a greater number of classic electrostatic bonds located in the right places to prevent the unfolding of the chain.[4] These discoveries were a bit surprising. So, weak bonds are not all that weak: taken individually they are very vulnerable, but if there are a lot of them, and they are well placed, they endow the proteins with unexpected solidity. There is strength in numbers.

An increase in the number of ionic bonds on the surface of proteins of microbes from hell also explains why hyperthermophiles cannot live in classic temperatures and why they stop dividing below 50 or 60°C (even, for some, below 80°C). At a low temperature, their proteins begin to freeze through an overdose of weak bonds. As for enzymes, they usually become too rigid to interact effectively with their partners and thus to play their role as catalysts. They must be activated by temperature to function normally. Note that there are exceptions and that some enzymes from hyperthermophiles can still work at low temperatures (although they are always much more active at high temperatures).

Hell: A Godsend for Science and Industry

As I recalled in the introduction to this work, microbes from hell owe part of their current celebrity to the use of one of their enzymes—DNA polymerase of *Thermus aquaticus*—during a polymerase chain reaction (PCR) to amplify DNA, a pillar of current molecular biology.[5] This enzyme is capable of enduring spikes in temperature to 90°C employed by this amplification method. More generally, in many industrial or biotechnological applications, classic enzymes are replaced with isolated proteins of hyperthermophile microbes (sometimes called extremenzymes), all the more interesting since those enzymes are active and stable not just at high temperatures: they are often also much more resistant to solvents or other chemical products used in industrial processes.[6]

Also, through the study of proteins of hyperthermophilic microbes, many researchers hope to learn how to modify a classic protein of industrial interest to make it more robust. This is another of the many examples in which basic research and applied research are closely connected. Starting with a question that might appear rather esoteric—Why doesn't the protoplasm of "plant" microbes from Yellowstone coagulate?—we end up with enzymes that will work for us endlessly, day and night, in a modern industrial hell.

The proteins of hyperthermophiles present another advantage, this time for biochemists. We don't really know why, but they are generally much easier to analyze through X-ray diffraction than are the proteins of mesophilic organisms: perhaps due to their rigidity at low temperatures, they form more beautiful crystals, and those diffract better. A scientist studying a protein present in many organisms, including hyperthermophiles, might choose to determine the protein's structure in the latter, since, as we have seen, the shape of a hyperthermophile protein and of one of its mesophilic counterparts is often very close. Thus it will be possible to extrapolate the results obtained from the hyperthermophile protein onto one's preferred protein—a human protein, for example—which is all the more interesting in that human proteins are in general rather fragile and difficult to crystallize. Here again we have an unexpected outcome from the curiosity of scientists who have looked at the hot springs of Yellowstone or Vulcano, or in the great ocean depths. The structure of certain human proteins associated with cancer can today be determined thanks to microbes from hell.[7]

The DNA Double Helix: More Weak Bonds

We will now look at the fate of another essential macromolecule of the living organism—DNA—at high temperatures. James Watson and Francis Crick discovered the structure of DNA in 1953; today almost everyone is familiar with it.[8] But do you know that this beautiful double helix also owes its existence to weak bonds that enable the connection of the two "simple" helices that comprise it? The DNA molecule is formed of two strands winding around each other. Each of them is a polymer of four nucleotides, and each is formed from the combination of a nitrogen base, deoxyribose (a sugar), and a phosphoric acid linked by strong covalent bonds (figure 3.6). Each of the four nucleotides has symbols that are the first letter of the base: A for adenine, T for thymine, G for guanine, and C for cytosine. In genes, the order of these nucleotides, the sequence, constitutes the genetic message: as we will see a bit later, this sequence determines the linking of the amino acids of different proteins, and therefore their structure and their function.

The genetic message carried by one of the two strands is the complement of the message carried by the other strand; it is this complementarity that enables the reproduction of the message from one generation to another. Notably, this complementarity is determined by the weak bonds of hydrogen type, which are established between the bases located opposite each other on the two strands of DNA. Indeed, the two strands do not carry the same series of nucleotides, but two complementary sequences that are mirror images of each other. For example, let's take a small fragment of human genetic code on one strand: TTCGCCAT. We note that it is mirrored on the other strand by a sequence AAGCGGTA, following a very simple rule: an A is always paired with a T, and a G is always paired with a C (figure 3.2). This is the pairing of the bases (we speak of base pairs AT or GC). Why does DNA follow this rule? Precisely because only the molecules A and T on the one hand, and G and C on the other, can "recognize," "touch," associate with each other, owing to the complementary hydrogen bonds between the atoms of one base and those of the other (figure 3.2). If we pair an A with a G or a C with a T, they don't "stick": there is no affinity between the two bases because they are incapable of bringing their electron clouds together to create stable hydrogen bonds. The specific complementarity between bases of DNA (and RNA) is the real "secret of life" revealed by the work of Watson and Crick, possibly one of the most important scientific discoveries of the millennium.

During cellular division, when the two strands of the double helix separate, each of them can serve as a matrix for the formation of a new strand whose sequence is complementary to that of the first, because stable associations can be established only between nucleotides carrying complementary bases. These are specialized enzymes—the famous DNA polymerases, which we've already discussed regarding PCR—that carry out the operation. From the sequence TTCGCCAT of one of the two strands, they create the sequence AAGCGGTA and reconstitute a double helix. Reciprocally, from the sequence AAGCGGTA of the second strand, they synthesize the sequence

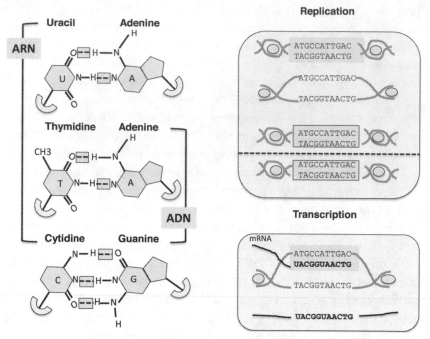

FIGURE 3.2. The language of nucleotides. Left panel: the AU, AT, and GC base pairs. The formation of hydrogen bonds between the atoms of bases is indicated by gray squares and dotted lines. Note that bases A and U(T) are linked by two hydrogen bonds, whereas bases G and C are linked by three hydrogen bonds. Right panel: replication and transcription of a nucleotide sequence. Hydrogen bonds between the atoms of bases are indicated by gray rectangles and circles. DNA is depicted with thin letters and gray lines, while RNA is depicted with black letters and thick lines. During transcription, the genetic message written in nucleotides is transcribed from DNA to RNA. The message is later translated into an amino acid message (proteins) during translation (change in language).

TTCGCCAT. As we can see in figure 3.2, we thus obtain two double helices identical to the first: this is the replication of DNA. This mechanism enables the DNA molecule to perpetuate the genetic message characteristic of a species from generation to generation.

So one of the mysteries of heredity is solved: it's a story of weak bonds. When Watson and Crick discovered the rules of base pairing between AT and GC in March 1953, they immediately understood that they had uncovered one of the secrets of life. It is said, moreover, that Crick immediately went to announce his discovery at the Eagle, his favorite pub in Cambridge.

Also important, the weak bonds between bases not only enable the recognition but also stabilize the entire DNA molecule by keeping the two strands tightly stuck together by a huge number of hydrogen bonds. Furthermore, the hydrophobic cores of each base pair (the atom rings) interact with the ones above it and the ones below via hydrophobic bonds, making the double helix even more stable. This is why it takes energy and enzymes to open the double helix. These are helicases (which break helices): enzymes using ATP that will do the work. We also saw in chapter 2 that the gyrase will facilitate this work in bacteria by activating the DNA. The "unscrewing of the DNA" introduced by this enzyme when it makes a negative superhelix will facilitate the breaking of hydrogen bonds between the two strands of the helix.

The weak bonds between the complementary bases are crucial for the double helix of DNA, so aren't we dealing here with the Achilles' heel of microbes from hell? Indeed, if we heat a linear DNA double helix (like the ones that make up our chromosomes) in a test tube, the hydrogen bonds gradually break and, around 70 to 80°C, the two helices abruptly separate from each other: this is DNA denaturation. This is why biologists have long thought that life could not exist above 70°C.

THE DOUBLE HELIX IN HELL: TOPOLOGY IS EVERYTHING

How do hyperthermophile microbes avoid the denaturation of their DNA? Is their DNA different from that of the common mortal—is it a "super DNA"? As with proteins, the response is no. The DNA of hyperthermophilic microbes is also formed by the intertwining of two helices through hydrogen bonds. So how does it survive at high temperatures? Biologists originally thought that the DNA of these uncommon microbes should endure high heat owing to a predicted high proportion of GC pairs. Indeed, the G and C bases are linked by three hydrogen bonds, versus only two for the A and T bases (figure 3.2). The more GC pairs the DNA contains, the higher the temperature necessary to denature the molecule (i.e., to separate the two strands). However, this prediction turned out to be wrong. The DNA of hyperthermophiles is far from being exceptionally rich in GC pairs. Just like in mesophilic organisms, the proportion varies greatly from one species to another; the DNA of microbes from hell can even be very rich in AT pairs. The chromosome of *Sulfolobus*, for example, has close to 65% AT pairs.

This observation has perplexed many researchers. But it wasn't really a problem. We discovered this by heating a small bacterial plasmid (corresponding to a circular double-stranded DNA molecule in which the two strands are rolled continuously around each other) in a test tube in the laboratory. Évelyne Marguet, whose underwater adventures we followed in chapter 2, carried out the experiment in 1994 in a laboratory at the CEA (Commissariat à l'énergie atomique et aux énergies alternatives—Atomic Energy and Alternative Energies Commission) in Saclay. The apparatus she used enabled her to measure the respective quantities of single-stranded or double-helix

DNA (these two forms absorb ultraviolet light differently). Évelyne incubated a solution of plasmids in this apparatus, which allowed the temperature of the solution to be gradually increased while measuring the absorption of the ultraviolet (UV) rays to deduce the relationship between the quantity of DNA with single and double strands. The results were spectacular: a double-stranded circular DNA could conserve a double-helix structure at least up to 107°C (the maximum temperature our equipment could reach).[9]

At this extreme temperature, we saw no modification in the absorption of UV. However, if we cut the plasmid using a restriction enzyme to make it straight, the double-stranded DNA disappeared abruptly in favor of the single-stranded DNA at 78°C; the two strands of the helix had separated, and the DNA had been denatured (figure 3.3).[10]

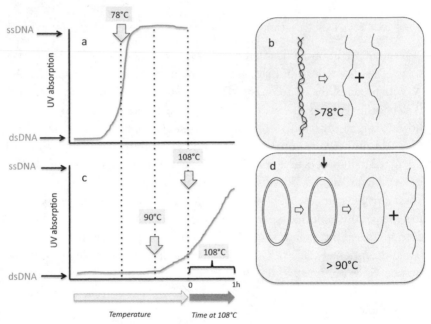

FIGURE 3.3. Thermodenaturation and thermodegradation of DNA. A, C, two experiments in which the fate of a linear DNA (A) or a circular topologically close DNA (B) at high temperatures is followed increasing in UV light absorption at 260 nm. In both experiments, the temperature was progressively increased from 50 to 108°C. A, the separation of the two strands occurs abruptly around 78°C, producing two linear single-stranded molecules (B). C, only a small portion of the DNA is denatured at 108°C. The DNA is then incubated for another hour at 108°C. DNA denaturation takes place progressively (not abruptly, as in A) because denaturation is a consequence of the progressive thermodegradation of DNA. Thermodegradation starts with the introduction of a break in one of the two strands (arrow in D). The break removes the topological constraints that prevented denaturation of topologically close circular DNA, leading to the instant denaturation of the molecule with production of one circular and one linear strand (D) (from Marguet and Forterre, 1994).

How can we explain these results? Outside the laboratory, what is their biological significance? In a small molecule of circular DNA, it is impossible to separate the two interlaced strands without breaking them. Even if through heating one suppresses all the weak bonds, the denatured DNA, in these conditions, forms a very specific structure that researchers call a random coil, in which the two strands of DNA are tangled randomly as many times as they were harmoniously folded around each other in Watson and Crick's double helix. In the experiment we carried out with Évelyne at Saclay, no statistical coil appeared while carrying a circular plasmid to 107°C since we didn't detect the appearance of single-stranded DNA.

We can easily carry out this type of experiment at home by winding together two pieces of string (A and B), then attaching them by each of their ends (an end of A with its other end, the same for B). It then becomes impossible to separate the two pieces of string, unless one of them is cut: A and B are connected by a topological bond (from the Greek *topos*, "place"). It is not a chemical bond, since the atoms that make up the two strings do not establish any particular relationship between them. Another example of a topological bond is the one connecting a key with a key ring. Here too there is no chemical bond between them: they are not glued together, but they cannot be separated unless you use the gap in the key ring, which corresponds to a break. Similarly, you cannot separate the two helices of a small circular DNA molecule unless you break one of the two to enable the other to pass through. And so circular DNA will be much more resistant to denaturation through temperature than linear DNA, because the formation of the statistical coil following the dissociation of weak interactions among its bases implies a lot of energy. In such a structure, the two DNA strands, although separated physically, are forced to remain very close to each other, in perpetual agitation, which leads to inopportune connections. In particular, the negative charges of the phosphorous groups that form their skeleton collide each time they meet, which nature doesn't like at all: the electrical charges of the same sign have a tendency to repel each other.

In the experiment we carried out at the CEA, we observed that the double helix of a small circular DNA is stable in a test tube at temperatures typical of places where hyperthermophiles exist, but is this observation relevant from a biological point of view? Yes, for several reasons. First, just like plasmids, the chromosomes of bacteria and archaea are almost all circular. Also, in all organisms, whether mesophilic or hyperthermophilic, many proteins are fixed on the DNA double helix, so that the two strands cannot freely unwind. The DNA in the cell must thus behave exactly like the small circular plasmid of our experiment; it is naturally resistant to denaturation. This explains why the genomes of microbes from hell are not richer in GC bases than those of other living beings.

Let's note that if the existence of topological links enables the DNA of microbes from hell to remain stable above 100°C, these same links create serious problems for all cells when the DNA must be replicated or transcribed. When he learned about the Watson and Crick model, Nobel prize winner Max Delbrück wrote to them, "I am willing to bet that the coiling of the chains around each other[11] in your structure is radically wrong," because he could not see how the cell can manage to separate the two strands of huge DNA chain if they are intertwined all along the molecule. Indeed, each time a turn of the double helix is opened during DNA replication, which corresponds to a distance of around ten pairs of bases, a topological link must be removed. Otherwise, the rest of the DNA molecule will fold on itself to compensate for the local unwinding of the two strands, forming superhelices that would rapidly block the machinery replicating DNA. This is what scientists used to call "the topological problem" linked to DNA replication. Watson and Crick were more optimistic than Delbrück. They wrote, "As in our model the two chains are wrapped around each others [sic], it is essential that they could be unwound to be separated . . . although it is difficult at the moment to see how these processes occur without everything getting tangle [sic], we do not feel that this objection will be insuperable."[12] The topological problems that bothered these early pioneers of molecular biology are resolved by the enzymes I have been studying for so long: DNA topoisomerases.[13]

MAGICIAN ENZYMES: DNA TOPOISOMERASES

Why that name? DNA topoisomerases have the unique ability to modify the topology of DNA molecules. For instance, they can transform a circular DNA molecule in the supercoiled shape—with a superhelix—into a DNA molecule in the relaxed form—without superhelix, or on the contrary turn superhelices into a previously relaxed DNA, as in the case of gyrase and reverse gyrase, discussed in the previous chapter. The substrate and products of these reactions catalyzed by DNA topoisomerases have the same size and sequences before and after the reaction. They only differ by the number of times the two DNA strands wrap around each other, a value called the topological linking number.[14] DNA molecules that differ only by their linking number are called DNA topoisomers; enzymes that can transform one type of topoisomer into another are called DNA topoisomerases. The person who discovered these enzymes, James Wang,[15] called them "magicians" precisely because they modify only a mathematical property of their substrate, the topological linking number (Lk).

To carry out their tasks, the topoisomerases should transiently cut one or two strands of the DNA molecule, which is the only way to modify the degree of intertwining between the two DNA strands that are topologically linked. Some of them make a transient cut in one strand of the double helix and make the other strand pass through the cut; they are called type I DNA topoisomerases. Others transiently break the two strands of the double helix at once and make another double helix (from the same molecule or from another molecule)[16] pass through this cut; they are called type II DNA topoisomerases (Figure 3.4).

Once the amount of coiling is modified according to the needs of the cell, the DNA topoisomerases close the cut or cuts they introduced in the DNA. The molecule is again intact; only the number of topological links has been modified.

Beside relaxing or supercoiling DNA molecules, DNA topoisomerases can untie knots in a DNA molecule, or even coil/uncoil the circular rings of DNA, plasmids, or chromosomes. Thus during cell division, type II DNA topoisomerases are indispensable for the separation of the two chromosomes that are systematically wound around each other at the end of the replication process. They can cut the double helix present in one chromosome, forming

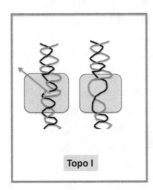 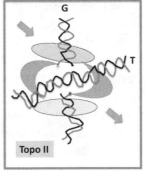

FIGURE 3.4. Type I and type II DNA topoisomerases. Type I DNA topoisomerases are monomeric proteins that produce transient single-stranded breaks into DNA and transport one DNA strand across the other, increasing or decreasing the linking number between the two DNA strands by steps of one. Type II DNA topoisomerases produce transient double-stranded breaks into DNA and transport one double helix (the transported segment T) across another one (the gate segment G). Type II DNA topoisomerases of bacteria and archaea are tetrameric proteins composed of two copies of two different subunits (*light and dark gray*). During the reaction, both type I and II DNA topoisomerases form transient covalent links with DNA (*small circles*). Some antitumoral drugs and antibiotics stabilize these links, producing poisons that can kill cells.

a double-stranded break, and force the sister chromosome to pass through the break. We will see in chapter 5 that the characterization of type II DNA topoisomerases found in archaea has led to surprising discoveries related to the sex of eukaryotes (including ours) and to the size of plants!

Note that DNA topoisomerases are not mere curiosities for biologists fascinated by DNA. In bacteria they are the target of commonly used molecules, such as antibiotics. Shortly after the 9/11 terrorist attacks you might have heard about Cipro,[17] an antibiotic used to fight anthrax, or carbon illness: this drug inhibits the famous DNA gyrase of bacteria (figure 2.1), which is a type II DNA topoisomerase and is indispensable to their survival. This antibiotic has no effect on humans, since, if you remember chapter 2, we don't have a gyrase. In humans, other type I and II DNA topoisomerases are the targets of several anticancer drugs used in chemotherapy.[18] If one blocks their activity, the targeted cell will have difficulty surviving. Cancer cells have many DNA topoisomerases, which enables them to divide rapidly; they are particularly sensitive to the inhibitors of these enzymes.

Because of their properties, DNA topoisomerases are capable not only of removing topological bonds but also of adding them. This explains, for example, why the reverse gyrase, a type I DNA topoisomerase, is capable of producing positive superturns. Each of these superturns corresponds to a topological bond that is added to those corresponding to normal turns of the double helix (see note 14). This increase in the number of topological bonds must stabilize the DNA molecule at high temperatures. At the moment of its discovery in 1984, it was logical to think that the role of the reverse gyrase in hyperthermophilic microbes was to protect the DNA of microbes from hell against thermal denaturation. But we have just seen that a bacterial plasmid, overcoiled negatively by the gyrase, is not denatured, at least not up to 107°C. Three years later, Olivier Guipaud, a student in our lab at Orsay, and Évelyne Marguet showed that an isolated plasmid of the hyperthermophilic bacterium *Thermotoga maritima*, which contains both the gyrase and the reverse gyrase, was negatively overcoiled.[19] Thus we had confirmation that one could live in hell with an overcoiled DNA, just as in heaven. Later, Purificación López-García, a Spanish postdoctoral student who was working with us at Orsay, made the same observation on the plasmid of a hyperthermophilic archaea, *Archaeoglobus fulgidus*, which also had the gyrase and the reverse gyrase.[20] In passing, in this case it is the gyrase that wins.

Discovering hyperthermophiles whose DNA is negatively supercoiled (thus more easy to denature; see figure 2.1) confirms, as revealed by Évelyne's experiment at the CEA, that topological constraints are enough to prevent the denaturing of DNA at high temperatures. But in that case, what does

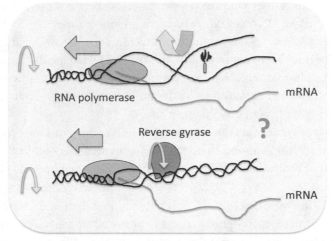

FIGURE 3.5. Hypothetical model for reverse gyrase acting as a renaturase. Top drawing: during transcription, the movement of RNA polymerase along DNA triggers rotation of the DNA double helix, overwinding DNA in front of the RNA polymerase and underwinding DNA behind. Underwinding can lead to denaturation of the double helix at high temperatures. Bottom drawing: reverse gyrase restore normal DNA winding, preventing denaturation behind the moving RNA polymerase.

the reverse gyrase really do? Some think that the reverse gyrase could be a "renaturase," an enzyme that would facilitate the reshaping of the double he lices at high temperatures after their separation at the time of replication and transcription (figure 3.5).

Is that hypothesis correct? At present, we're not sure. In any case, we will see in chapter 5 that the reverse gyrase is certainly indispensable for life at high temperatures, and that its evolutive history provides data that will prove decisively enlightening regarding the origin of hyperthermophiles. This makes the fact that we don't really understand its use all the more frustrating.

REPAIRS IN HELL

In the experiment we carried out with Évelyne at Saclay, a negatively super-coiled plasmid remained in the form of a double helix up to 107°C. However, if the incubation to 107°C continued for several minutes, up to one hour, the DNA was finally denatured. We observed the gradual appearance of single-stranded DNA by following the absorption of UV rays; after some time, the DNA was entirely single stranded[21] (figure 3.3). What happened? This observation reflects the gradual introduction of breaks in one of the two strands of DNA.[22] The breaks observed correspond to the rupture of the (not so strong) covalent bonds that link nucleotides. This is a spontaneous chemical reaction whose speed increases with the temperature. As soon as a break occurs in a plasmid, the two DNA strands, which are no longer topologically linked to each other, separate instantaneously, producing a circular strand and a linear strand (the one that has been broken; figure 3.3). During incubation at 107°C, millions of plasmid molecules present in the test tube will gradually break, one after the other, explaining why the proportion of single-stranded DNA increases regularly with time (figure 3.3).

DNA breakage occurs not only in vitro but also in vivo. However, in 1995 an American team showed that DNA in the hyperthermophile *Pyrococcus furiosus* is about twenty times more resistant to thermal breakage than the bacterium *Escherichia coli*. They observed six times fewer breaks in the *P. furiosus* cell after exposure to 110°C for thirty minutes than in *E. coli* at 95°C.[23]

How do microbes from hell manage to protect their genetic material from thermal degradation?[24] I have shown with Évelyne that thermal degradation of DNA can be slowed by adding salt—for example, kitchen salt, sodium chloride—to the test tube.[25] It is thus likely that high salt concentration in the cytoplasm of hyperthermophiles may be a trick, somehow allowing these organisms to prevent DNA degradation at high temperatures. Unfortunately, we don't really know the intracellular salt concentration in most hyperthermophiles.

The resistance of microbes from hell to DNA damage induced by heat could be related to another observation made some twenty years ago by scientists from the former Soviet Union: hyperthermophile archaea are very resistant to radioactive rays.[26] They tolerate doses of radiation a hundred times higher than those that killed the inhabitants of Hiroshima or the workers in Chernobyl. This observation was then repeated and confirmed by a French biologist who emigrated to the United States, Jocelyne DiRuggiero, and by a student in our lab at Orsay, Emmanuelle Gérard.[27] A student in Daniel Prieur's lab in Brest, Edmond Jolivet, whom you met on the *Atalante* during the AMISTAD expedition, then isolated a very radioresistant hyperthermophile archaea by subjecting samples from the smokers brought back from the depths of the Sea of Cortez, off the coast of Mexico, to very strong doses of radioactivity. He came to work with us at Orsay, where he showed that this archaea, which he christened *Thermococcus gammatolerans* ("the thermal shell tolerant to gamma rays"),[28] was almost as radioresistant as the famous *Deinococcus radiodurans* bacterium, isolated in the United States in the 1950s from a sterilized jelly jar by exposing it to radioactive gamma rays.

The analysis of the DNA of hyperthermophiles just after irradiation by gamma rays (a very violent radioactive bombardment) has shown that their

DNA was not inherently protected against rays. The chromosome of irradiated hyperthermophiles is broken into dozens of linear fragments. It is thus no longer topologically closed. However, as soon as microbes are replaced in a culture medium at 90°C, instead of observing an instantaneous denaturation of these bits of DNA, we see that they are reconnected, after an hour, to form a single large circular chromosome. How is this possible? Luckily for them, hyperthermophiles—just like other organisms—have specialized enzymes whose role is to repair DNA.[29] But are they sufficient to fulfill the requirement for a DNA temperature at 100°C, or are there DNA reparation mechanisms specific to microbes from hell?

To advance in our knowledge of the resistance of *Thermococcus* to radiation, in the early 2000s I obtained grants from the CNRS to sequence the genome of *Thermococcus gammatolerans*.[30] This enabled me to fund the research of Fabrice Confalonieri, a research assistant working at Orsay who wanted to start his own team at the Institut de Génétique et Microbiologie. Among other things, I wanted to thank him for the important work he had achieved on the reverse gyrase, which we will talk about more in the next chapter. Also, he had a lot of expertise in the analysis of genomes. For the moment, we haven't really found any DNA reparation mechanisms specific to hyperthermophiles by analyzing the genome of *T. gammatolerans* or other microbes of hell resistant to radiation.

POISONS IN HELL

Beside radiation, hyperthermophiles are especially resistant to other stresses that can damage their DNA or other macromolecules, such as proteins. This is most likely owing to the unusual environments where they live, which are full of poisonous substances. For instance, hydrothermal vents are very rich in heavy metals, such as cadmium, zinc, cobalt, and nickel. Christian Jeanthon, who led the AMISTAD expedition, has shown that *Thermococcales* are indeed highly resistant to cadmium and zinc.[31]

In 2004 I obtained a grant from the CEA program ToxNuc (nuclear toxicology) to study the resistance of *T. gammatolerans* to heavy metals. With this money, the team of Confalonieri, who had since become a professor, was able to show that *T. gammatolerans* was indeed very resistant to cobalt, cadmium, and zinc. They have studied how exposure to cadmium modifies gene expression and found that about 5% of the around 2000 *T. gammatolerans* genes change their expression following exposure to cadmium.[32] They observed in particular the induction of several genes encoding proteins (transporter) that can potentially expel damaged compounds from the cell or proteins that can recruit iron for the cell, possibly to compensate for damage induced in iron-containing proteins of the cells.

Is it possible that the membrane vesicles that we are studying now at Orsay are themselves involved in these processes of detoxification? Collaborating with François Guyot, a geochemist from the National Museum of Natural History in Paris, Aurore Gorlas has shown that these vesicles might be used by cells to detoxify excess sulfur,[33] and that they have the ability to accumulate minerals added to the external environment and even to catalyze reactions among these minerals.

Once again, we see that microbes from hell are not small, fragile beings that barely manage to survive in their extreme environments, but marvels of adaptation that can successfully confront several extreme conditions at the same time: high temperatures, acidity (for thermoacidophiles), and the presence of heavy metals. But let's go back to nucleic acids. We have just talked about DNA; we will now turn to its ancestor, RNA.

RNA MOLECULES: TO KNOW THEM IS TO LIKE THEM

In the family of nucleic acids, everyone knows about DNA and its beautiful and photogenic double helix, but its close relative, RNA, its ancestor, remains almost unknown by the media, as we will soon see. RNA are, however, particularly important macromolecules. In the next chapter we will learn that RNA probably appeared on Earth well before DNA, and probably preceded proteins as we know them today.

RNA looks a lot like DNA: it is a macromolecule made up of a long series of four different nucleotides attached to each other by covalent bonds (figure 3.2). These nucleotides nonetheless present two important specific characteristics that differ from DNA. On the one hand, the thymine base (T) is replaced by the uracil base (U), which, like T, pairs specifically with the adenine base (A). On the other hand, the sugar of these nucleotides is not deoxyribose, but ribose (RNA is ribonucleic acid). Furthermore, this macromolecule appears very rarely in the form of a double helix, except in certain viruses whose genome is made up of double-stranded RNA (such as rotaviruses responsible for some gastroenteritis). In general, RNA molecules are formed by a single chain, most of the time much shorter than the chains of DNA: RNA strands include only a few dozen to several thousand nucleotides. They sometimes fold over themselves to form minihelices maintained by AU and GC pairings (figure 3.6c).

RNA molecules are involved in a multitude of functions vital for the cells

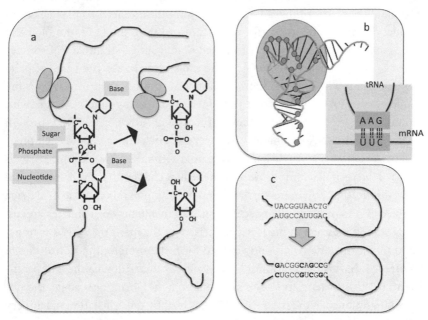

FIGURE 3.6. Thermodegradation and protection of RNA at high temperatures. *A*, cleavage of RNA at high temperature is triggered by the reactive oxygen (*in gray*) of the ribose by attacking the phosphate linking two nucleotides. This prevents ribosomes from completing translation of the messenger RNA. *B*, schematic view of a tRNA molecule. The RNA strand (*gray line*) folds onto itself to form two minihelices (*black lines*). The gray circles indicate positions in which the bases (A, U, G, or C) are often chemically modified in hyperthermophiles to protect rRNA against heat damage. *C*, an RNA minihelix from a mesophile (*top*) and a hyperthermophile (*bottom*). The latter contain a higher percentage of GC bases (*bold*), increasing the number of hydrogen bonds stabilizing the helix.

of all living beings. In particular, they play a major role in the transcription and translation of the genetic message carried by DNA (figure 1.3). We will discover that the accomplishment of these two essential processes raises important problems at the temperatures in which hyperthermophiles live.

FROM NUCLEOTIDES TO AMINO ACIDS

To truly understand the obstacles that microbes from hell must overcome, we need to remember the essential characteristics of the processes through which a cell synthesizes a protein from the message that is conveyed on the DNA by the nucleotide sequence of a gene.

This message is first faithfully retranscribed on an RNA molecule called a messenger RNA (figure 3.2). We speak of *transcription* because the language remains the same—that of nucleotides—but the support changes: we go from DNA to RNA. To remember this term, we can imagine a Coptic monk in the

Middle Ages transcribing a Greek text from an old piece of papyrus onto a sheet of parchment. To perform the transcription, the DNA double helix opens and one of the strands is recopied onto a single-stranded RNA molecule of complementary sequence—the messenger RNA. The rules of pairing associated with transcription are identical to those that preside over the replication of DNA, except that thymine is replaced by uracil: A pairs with U and G pairs with C. The synthesis of messenger RNA is achieved through an enzyme, the RNA polymerase, which adds nucleotides behind each other by covalent bonds so their bases are paired—thanks to weak bonds—with those of the recopied strand. We saw in the first chapter that Zillig was a great specialist in RNA polymerases; the characterization of this enzyme in archaea enabled him to prove an unexpected bond of kinship between archaea and us.

Next, the message in nucleotides carried by the messenger RNA is translated into a sequence of amino acids to form the protein: this is *translation* (figure 1.3). In this case, our monk translates his text, still on a sheet of parchment, but this time from Greek into Old French. During this stage, the nucleotide sequence of the messenger RNA is read by the translation machinery as if it were a text composed of three-letter words. In this sequence, each group of three nucleotides (called a triplet or codon) corresponds to one of the twenty amino acids present in proteins. This correspondence between codons and amino acids constitutes the genetic code. It is indeed a code, because it enables the translation of the language of nucleic acids into the language of proteins.

The translation of the messenger RNA depends on two other types of RNA: transfer RNA (figure 3.6b) and ribosomal RNA (see figure 4.2 in the next chapter). We encountered ribosomal RNA in chapter 1. It is the principal component of ribosomes, those complex structures in which messenger RNA is translated into proteins. The transfer RNA are the smart molecules that transfer the amino acids to the ribosomes according to the message carried by messenger RNA. The amino acid is attached to one extremity of the transfer RNA by a covalent bond, whereas the other extremity of the transfer RNA, called the anticodon, is bound to the codon corresponding to that amino acid, following the genetic code. Transfer RNA thus plays the role of adapter. It is fixed onto the messenger RNA on the level of the codon corresponding to the amino acid being transported, through weak bonds that are established between the bases of the codon and the anticodon following the classical AU/GC rule (figure 3.6b).

In very precise locations on the ribosome, two transfer RNA, each in charge of their amino acids, are paired with two successive codons on the messenger RNA (let's call the two amino acids A1 and A2, according to their position in the protein sequence). The RNA of the large subunit of the ribosome then catalyzes the formation of the (strong) peptide bond between these two amino acids. Once the peptide bond has been made, the transfer RNA that was charged with A1 is removed from the ribosome, which enables a third transfer RNA to settle, bringing the next amino acid (A3) that will be linked to A2, and so on, until the protein coded by the initial gene is entirely synthesized.[34]

I just mentioned that, in this process, the formation of the peptide bond is catalyzed by one of the two ribosomal RNA. No, you are not mistaken: it is not a protein but RNA that catalyzes this biochemical reaction. We will discuss this very important point again soon when we talk about RNA enzymes called ribozymes (not to be confused with ribosomes).

MORE WEAK BONDS CENTRAL TO LIFE

The first key stage of the translation process is that which guarantees a faithful deciphering of the message carried by the messenger RNA. It is based on the association of the right amino acid with the right transfer RNA. This reaction is catalyzed by enzymes—one for each amino acid—that bind a given transfer RNA with the amino acid corresponding to its anticodon, without making the slightest error.[35] These proteins must recognize very precisely the exact shape of each transfer RNA because it is they that indicate to which amino acid it must be associated. The three-dimensional structure of transfer RNA is thus a determining element of the faithfulness of the translation.

The second crucial moment of translation is that which conditions the synthesis of the protein: the formation of the peptide bond between amino acids by the ribosome. In this case, it is the three-dimensional structure of the ribosomal RNA that is essential, because it is one of the ribosomal RNA that, like an enzyme, catalyzes the reaction (the evolutionary significance of this fact is fundamental; we will discuss this in the next chapter); just as with proteins, the functioning of RNA enzymes depends a great deal on their shape in space.

The spatial structure of transfer RNA and ribosomal RNA is thus central to life. These molecules formed by a single strand are able to fold on themselves like proteins to form relatively compact structures, with, on their surface, one or several sites that can recognize a substrate, or contain atoms capable of catalyzing a chemical reaction, or both. The folding in space of a strand of RNA results in the establishment of a great number of weak bonds between the atoms of nucleotides located at distant points, mimicking the folding of the polypeptide chain in proteins via the establishment of weak bonds between amino acids located at distant points. In particular, as I've already mentioned, minihelices can be formed between complementary sequences carried by different regions of the same strand. For example, transfer RNA is made up of four minihelices folded on themselves, with the whole adopting a structure in the shape of an inverted L (figure 3.6b). The minihelices are stabilized by the pairing of A and U bases on the one hand, and G and C bases on the other, which are in contact through two and three hydrogen bonds respectively (figure 3.6c).

Molecules structured by weak bonds and whose biological function is inseparably linked to the structure risk thermal denaturation. In other words, the characteristics of the RNA that we have just described tolerate infernal temperatures very poorly.

RNA IN HELL

Increased Protection against Thermal Denaturation

We have seen that DNA is stabilized at high temperatures by the topological constraints engendered by its circular shape. But transfer RNA and ribosomal RNA are linear molecules formed by a single strand, so they cannot form topological bonds, and their complex spatial structure should normally unfold as soon as the temperature exceeds 80°C. However, this RNA must actually function perfectly up to 110°C in hyperthermophiles. In fact, the biochemists who have studied transfer RNA of hyperthermophiles have shown that in a test tube, they tolerate temperatures higher than 100°C, whereas those of mesophiles denature around 80°C.[36] How is that possible? Comparing the transfer RNA and ribosomal RNA sequences of hyperthermophilic microbes with those of their mesophilic counterparts immediately provides an initial response: in hyperthermophiles, AU base pairs of the minihelices are almost systematically replaced by GC base pairs, which increases their stability: as we have seen, guanine and cytosine are paired through three hydrogen bonds, whereas AU base pairs are linked through only two hydrogen bonds (figure 3.2). We find in transfer RNA and ribosomal RNA the correlation between high temperature and a wealth of GC bases that had been postulated, but not observed, in DNA. This confirms that the absence of correlation in DNA can be explained by the stabilization of this macromolecule through topological bonds.

But the wealth of GC base pairs is not enough to explain the extraordinary thermal resistance of the transfer RNA and the ribosomal RNA of microbes from hell. Another "trick" used by hyperthermophiles was discovered by an

American biologist, James McCloskey, in the early 1990s. It has been known for a long time that in mesophilic organisms, ribosomal and transfer RNA, apart from the four classic nucleotides, possess modified nucleotides containing supplementary groups of atoms, either on the base or on the ribose. By analyzing the transfer RNA of hyperthermophilic archaea using very sophisticated chemical techniques, McCloskey was able to show that the latter have many more modified nucleotides than their mesophilic counterparts (this corresponds to positions labeled with a grey circle in figure 3.6b).[37] The supplementary atoms of these modified monomers enable the formation of new weak bonds between distant regions of the molecule. These bonds serve in particular to reinforce the connection of minihelices in the final structure of the RNA molecule. These minihelices, truly "glued" to each other, resist the ambient agitation induced by high temperatures. Not only are there more modified nucleotides, but some of them are truly exclusive to hyperthermophiles: they are found nowhere else on the tree of life. Researchers have, moreover, observed that the transfer RNA of a given hyperthermophile becomes all the richer in modified monomers as the microbe is subjected to higher temperatures.[38]

A Reinforced Skeleton

A remarkable fact jumps out at us when we closely observe the transfer RNA and the ribosomal RNA of microbes from hell. Very often, one of the oxygen atoms of the sugar that forms the skeleton of the nucleotide—the ribose—carries an additional chemical group: the methyl group (one carbon atom combined with three hydrogen atoms: CH_3).

Let's take a little detour here to sing the praises of ribose: it is probably the most important molecule for all living beings that inhabit planet Earth, for it is ribose to which we owe our lives. I therefore encourage you to consider figure 3.7 with respectful attention.

Ribose owes its name to the philanthropy of the Rockefeller family, whose fortune helped finance many American research laboratories in the last century. This sugar (ose) was discovered at the beginning of the twentieth century at the Rockefeller Institute of Biochemistry. Ribose, whose skeleton is made up of five carbon atoms, is present not only in RNA but also in ATP (adenosine triphosphate), a small molecule that acts as fuel and is the energy source of all cells in the living world. It is also found in the nucleotides of DNA in a slightly modified form: deoxyribose. Deoxyribose is missing an oxygen atom—the one that is often methylated in the transfer RNA and ribosomal RNA of hyperthermophiles. Notably, the presence of an oxygen atom at

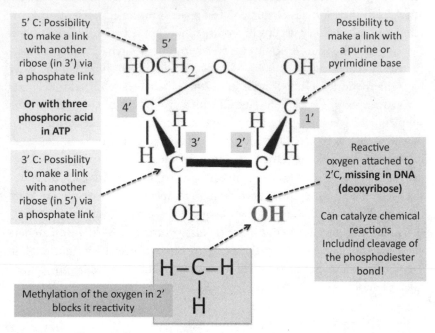

FIGURE 3.7. The magic ribose.

this position (called 2′ by biochemists)[39] is a major weak point of RNA at high temperatures. It's very important to understand why, in order to comprehend why the RNA molecule must be carefully protected in the heat of hell.

In RNA or DNA chains, nucleotides are connected by covalent bonds between ribose or deoxyribose of a nucleotide and the phosphoric acid of a neighboring nucleotide. These bonds form a "phosphate bridge" between oxygen atoms belonging to the sugars of two neighboring nucleotides (figure 3.6a). In RNA, the formation of this phosphate bridge leaves oxygen fixed on the carbon on 2′ of the ribose in a free state. This is a critical point because this atom is very reactive: it often intervenes, for example, in reactions catalyzed by some RNA molecules mimicking enzymes, the famous ribozymes. In fact, the absence of the oxygen fixed on the carbon on 2′ of the deoxyribose in DNA explains why there are no DNA enzymes!

But if the presence of a free 2′ oxygen in RNA provides this molecule with catalytic properties, the reactivity of this oxygen has a negative side because it causes the RNA molecule to be fragile. The oxygen atom possesses the annoying tendency of "attacking" the phosphorus atom that connects two neighboring nucleotides. This confrontation results in the rupture of the bond between the two monomers: the RNA strand breaks in two (figure 3.6a). The RNA is thus much less stable from a chemical point of view than is DNA. Attacking implies movement; thus as the temperature rises, the attacks become more frequent. RNA is therefore more fragile than DNA at high temperatures: RNA is a particularly thermolabile molecule.[40]

All the same, when the oxygen located on 2' of the ribose is connected by a covalent bond with a methyl group, its chemical reactivity can be neutralized. One can precisely observe in transfer and ribosome RNA of microbes from hell the presence of many riboses whose oxygen situated on 2' is blocked by a methyl group (CH_3).[41] As in the case of other modifications, these methyl groups can interact with the atoms of other nucleotides of the RNA chain by forming hydrophobic-type bonds to stabilize the structure of the RNA molecule. But I think they also play an important role in protecting the transfer RNA and the ribosomal RNA against thermal degradation, killing two birds with one stone. They must be present on the riboses located at strategic placements of the molecule, probably where they are most accessible to water molecules that might come to help in the attack of the bond between nucleotides by a reactive oxygen molecule. It is thus more than probable that the methylation of riboses plays an important protective role. All the same, to my knowledge, no one has yet tested this hypothesis. It must be said that microbiologists who are interested in microbes from hell are rarely RNA specialists; likewise, RNA specialists generally work on "model" organisms that live at the same temperatures as you and I. This is a field of research that remains to be explored.

There are two more questions to be asked concerning high-temperature RNA. First, how are the interactions between messenger RNA and transfer RNA—that is, between codons and anticodons—stabilized? They depend on weak bonds very sensitive to increases in temperature. The answer came from an analysis of genomes. Let's recall that if there are twenty amino acids, it is possible to write sixty-four different codons of three letters (4×4×4) with the four letters of the nucleotide alphabet of DNA and RNA. One of the sixty-four codons, AUG, signals the beginning of a gene (start codon); three others, UAA, UAG, and UGA, correspond to the end of the gene (termination codons). Therefore, several of the twenty amino acids can be coded through different codons. In general, the different codons corresponding to the same amino acid differ in the third letter. The genetic code is the same for microbes from hell as for all other living beings (we say it is universal), but thanks to the property that we have just pointed out, hyperthermophiles can modify its use. In general, for a given amino acid, when the first two letters of the codon are A or T (AA, AT, TA, or TT), hyperthermophiles systematically possess a

G or a C in the third position to stabilize the codon-anticodon interaction (figure 3.6b).[42] Living in high temperatures thus modifies the genetic message in a very subtle way on the level of the DNA itself (since the codons of the messenger RNA are recopied from the DNA). It must be noted that, as we have seen, this doesn't prevent some genomes of hyperthermophiles from being rich in A and T. In that case, the first two positions of the codons will necessarily be A and T.

The second question is this: What about the messenger RNA itself? How does it manage to tolerate high temperatures? Is it protected from thermal degradation by the methylation of the ribose, like the transfer RNA and the ribosomal RNA? This is not the case, and it wouldn't be imaginable in any event, because the enzymatic systems that fix the methyl groups at certain strategic points of the RNA molecules (on their surface or at the point of contact of two minihelices) must recognize the precise sequences within those molecules. By definition, the messenger RNA sequences are variable, since they reflect all the diversity of the genome.

Does messenger RNA even need to be stabilized? Not necessarily, no. In a bacterium like *Escherichia coli*, which lives in our intestines, we know that the messenger RNA are very unstable molecules that are degraded as soon as their message has been read by the ribosomes. This isn't a problem for the bacterial or archaeal cells, because new messenger RNA are continually produced (transcribed) from the DNA through RNA polymerases. On average, the half-life of the messenger RNA of a bacterium or an archaeon is a few minutes, which isn't long enough for them to be seriously affected by temperature.[43] The transfer RNA and the ribosomal RNA, on the other hand, live longer (biochemists usually call them stable RNA), which would explain why they must be protected against the effects of thermal degradation.

Why Aren't There Any Eukaryotes from Hell?

Here I raise a particularly important point. In chapter 1 we saw that only prokaryotes (bacteria and archaea) have managed to cross the barrier of 60°C: we don't find eukaryotes in hell. Why? An attractive hypothesis is that, unlike archaea and bacteria, their messenger RNA is not short-lived, but should be stable long enough to remain intact between the time of its synthesis in the nucleus and its translation in the cytoplasm (figure 1.3). As we have just seen, the coupling between transcription and translation that can take place only in prokaryotes is a powerful advantage for life at high temperatures. This coupling cannot take place in eukaryotes, since the RNA transcribed in the nucleus still contains the region corresponding to the noncoding introns sequences. This RNA should first be spliced into a messenger RNA that must then be transported from the nucleus to the cytoplasm to be taken in hand by the ribosomes. All these processes take time. The half-life of eukaryotes' messenger RNA can thus vary from a half hour to several days instead of minutes.[44] Would such a difference explain why there aren't any thermophile or hyperthermophile eukaryotes? The necessary stability in time of messenger RNA in eukaryotes could well be incompatible with life at high temperatures. I proposed this hypothesis in 1994 in an article published in the *Comptes rendus de l'Académie des sciences*.[45] Since then, as we saw in chapter 1, no one, not even Stetter, has been able to isolate a eukaryote capable of living permanently at more than 60°C, and no one has yet proposed other hypotheses to explain the absence of eukaryotes in hell. So would my hypothesis be correct? Perhaps. In any case, we will see in the next chapter that it is possible to combine it with a second hypothesis that would explain why archaea and bacteria have become prokaryotes.

The Maximum Temperature for Life: The Story of a Membrane

There is a final category of molecules of living organisms for which high heat poses a problem. They are not "giants," and they do not carry information in them, but they nevertheless play an essential role in the life of a cell. These are membrane lipids. Although I am talking about them at the end of the chapter, this is not to minimize their importance—quite the opposite. We will see that they play a crucial role in our story. Let me stress this point right away: they likely determine the maximum temperature at which life can exist on our planet.

THE CELL CYTOPLASMIC MEMBRANE

In the three domains of life, the cytoplasmic membrane that covers cells is always formed by a lipid layer with proteins inserted in it, facing either the extracellular space or the intracellular one (the cytoplasm). It is sometimes referred to as a "reverse sandwich," with the filling on the outside and the butter inside. The lipid layer is highly hydrophobic and plays a critical role in cellular life by preventing water, ions, and small hydrophilic molecules (such as ATP, amino acids, or nucleotides—precursors of DNA and RNA) to leak out of the cell. Because of the lipid layer, membranes are also impermeable to ions, in particular to sodium ($Na+$) ions and potassium ($K+$) ions, but also to protons.[46] It is very easy to appreciate the water phobia characteristic of the lipid layer. You only have to pour some oil—which is composed almost exclusively of such long carbonaceous chains—into a glass of water: the two liquids do not blend.

In most bacteria and all eukaryotes, the lipid layer is a "bilayer." This is because membrane lipids have a very specific structure: they are formed with a polar head (so they are often called polar lipids) and two tails. The polar head (open circles in figure 3.8a) corresponds to a region rich in hydrophilic chemical groups, often containing phosphorus, thus being able to form hydrogen or ionic bonds with water molecules, ions, or hydrophilic molecules.

In contrast, the two tails are long hydrophobic carbon chains (with from sixteen to twenty carbon atoms) able to form hydrophobic bonds with each other and with hydrophobic parts of proteins located within the membrane. The phospholipid bilayer is organized in such a way that the polar heads of the upper layer are oriented toward the extracellular space, and those of the lower layer are oriented toward the cytoplasm, two regions rich in water. Consequently, the hydrophobic tails point toward the interior of the membrane and form the zone from which water, ions, and other small hydrophilic molecules are excluded (grey zones in figure 3.8a). The connection of the two layers is maintained through van der Waals–type hydrophobic interactions. Although very weak individually, these hydrophobic interactions by their numbers ensure the cohesion of the great membrane edifices.

The existence of the lipid layer implies that the exchanges of molecules and ions between the cell and its environment must be controlled and regulated by specific proteins playing the role of transporters. Often, these latter form a cylinder, a membrane pore whose hydrophobic exterior surface enables immersion into lipids of the membrane (see the cylinder section in figure 3.8a). The hydrophilic interior of the cylinder creates a tunnel through which small polar molecules, recognized—thanks to weak bonds—by the transporter, pass through. The transporting of ions is particularly important because all cells have a very large intracellular concentration of potassium ions and a very weak concentration of sodium ions.[47] Amazingly, the opposite situation prevails in nature (in seawater, for example): sodium is very abundant and potassium very rare. Luckily, the impermeability of the cytoplasmic membrane prevents the sodium from the environment from freely invading our cells (figure 3.8a). In fact, we will see a bit later that specialized pumps enable sodium (or protons) to enter by specific channels to produce energy before rejecting them to the exterior (figure 3.8b).

Cells must also have effective transporters to let potassium enter (figure 3.8a). This is essential for the functioning of a great number of mechanisms necessary to life. In particular, a potassium ion is present on the level of the active site of the ribosome, where bonds between amino acids are

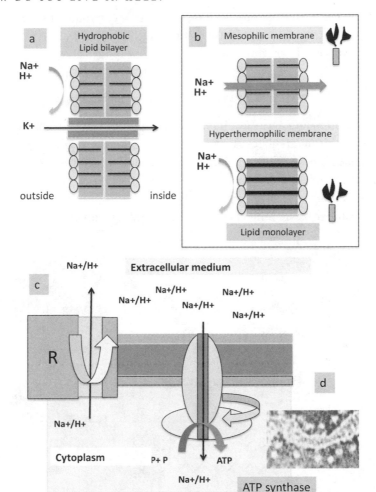

FIGURE 3.8. The wonderful world of cytoplasmic membranes. *A*, phospholipid bilayers membrane with a K+ transporter (pore) and monolayer membrane (*b, bottom*). *B*, comparison of bacterial/eukaryal mesophilic membrane (*top*) and archaeal hyperthermophilic membrane (*bottom*). At high temperatures, the archaeal membrane is not permeable to ions (Na+) and protons (H+) because of the thickness of the hydrophobic carbon (*thick lines*). *C*, cytoplasmic membrane with a respiration chain (R) leading to the excretion of ions and/or protons (H+). The ions/protons can equilibrate their concentrations inside and outside the cell by crossing the membrane via the ATP synthase pore. This movement triggers rotation of the internal subunit of the ATP synthase. This rotation produces the energy required for ATP synthesis. *D*, electron microscope picture of a mitochondrial membrane carrying "mushroom-like" ATP synthases.

formed.[48] This is a problem for the cell because potassium has little affinity for macromolecules, RNA or proteins, with which it must interact. The cells must thus spend a lot of energy to pump the little potassium present in the external environment so that its intracellular concentration is sufficient to enable it to be associated with molecular targets.[49] Why is potassium so important for the functioning of living organisms if it is so rare in the environment? We will see in chapter 5 how this puzzling observation led a German scientist to propose a new theory for the origin of life.

A Porous Epidermis at High Temperatures

How do lipid membranes react if they are exposed to very high temperatures? To answer this question, let's take a final small detour through biochemistry. Membrane lipids are glycerolipids, composite molecules formed through the combination of a glycerol molecule (the polar head) and two molecules of fatty acids (the two hydrophobic tails; figure 3.9).

The fatty acids are long chains of carbon ending with an acid carboxylic

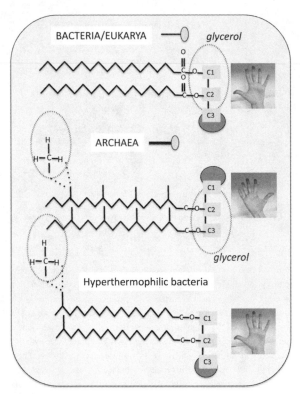

FIGURE 3.9. The great lipid divide. Glycerolipids from bacteria/eukarya and from archaea have different long carbon chains (fatty acids in bacteria/eukarya and isoprenol with multiple methyl groups in archaea) and linkage of these chains with glycerol (ester link—COO—in eukarya and most bacteria; ether link—CO—in archaea). They also have different stereochemistry (symbolized by left and right hands) with polar groups (*gray circles*) linking either to carbon 1 or 3 of the glycerol. Phospholipids from hyperthermophilic bacteria are typically bacterial in terms of stereochemistry, but the ester link can be replaced by an ether link and a methyl group is often present at the end of the carbon chain.

function (group COOH). It is through a covalent bond—called an ester bond—between the carboxylic acid functional group and an alcohol functional group (group OH) of glycerol that glycerolipids are formed. This ester bond is precisely one of the not so strong covalent bonds that are relatively fragile at high temperatures. In other words, there is an initial problem: glycerolipids plunged into water that is too hot risk losing their heads.

The second problem, which is even more serious, is that of membrane permeability. When the temperature rises, the properties of the lipid bilayer are modified (again, through the agitation of atoms) to such a degree that it becomes progressively permeable to protons and ions (figure 3.8b). For the cell, the consequences are dramatic. Not only does the "bad" sodium invade our cells and the "good" potassium escape, but the permeability to sodium and to protons will considerably slow down the fabrication of ATP, the major energy fuel of cells.

What is the connection between ATP and our membranes? Again, it's the miracle of electricity. When we eat and breathe, the end products of these activities are chemical reactions that occurred on the level of the cell membranes and generate a flow of electrons and therefore a local electrical current (figure 3.8c). The energy transported by this current enables specialized transporters to eject protons and sodium outside of the cell.[50] These reactions thus induce a disequilibrium: there are soon many more protons and sodium ions on the outside than the inside. If the membrane allowed protons and ions to diffuse freely, they would spontaneously hasten to get back into the cell. Fortunately, the membrane being impermeable to them, they cannot pass through except at the level of specific proteins, sort of inversed pumps, where their movement is at the origin of another electrical current (figure 3.8c). For the cell, the trick consists of recovering the energy associated with this current and converting it into chemical energy in the form of ATP.[51] This fascinating mechanism was discovered by the English Nobel laureate Peter Mitchell, who in 1961 proposed his "chemiosmotic theory."

The inversed pumps that achieve this feat are wonderful molecular machines called ATP synthases[52] (figure 3.8c, d). Under the effect of the flow of protons or sodium, these molecular microdynamos turn (literally, for one of their protein components, the rotating subunit) permanently on the internal surface of membranes, like the blades of a mill in water under the effect of the flow of the river. Through this process, the energy channeled enables the creation of a strong chemical bond that forms ATP from its precursor, ADP.[53] Of course, this fascinating mechanism can work only if the membrane is not permeable to sodium ions or protons. Otherwise, protons or sodium will bypass the ATP synthase and enter passively into the cell.

The problem for our microbes from hell is that, as we have seen, membrane permeability increases precisely with rising temperature. In the 1990s a team of Dutch biochemists, led by Wil Konings in Groningen, measured the temperature at which "normal" cell membranes become permeable to protons. The results were clear: starting at 70 to 80°C, they dangerously begin to allow protons to pass, and at 90 to 100°C, they become completely permeable (figure 3.8b).[54] In that case, the mechanism discovered by Mitchell cannot work anymore. Once ejected, the protons or sodium diffuse freely through the membrane without passing through the ATP synthases, and those enzymes will stop functioning.

In the 1970s, scientists showed that some bacteria whose membranes were rendered permeable to protons through drugs that make holes in their membranes can continue to live at low temperatures, although the proton motive force is abolished and their ATP synthases can no longer function.[55] They still manufacture a bit of ATP thanks to some metabolic paths, such as those used for the degradation of glucose: glycolysis. But it is likely that microbes from hell need to continuously produce ATP in large quantities to be able to constantly renew their macromolecules that have been damaged by exposure to high temperatures. In that case, the ATP synthases should be essential to hyperthermophiles.[56]

The Great Lipid Divide

We are now faced with a major question: If membranes become permeable at high temperatures, how can microbes from hell live between 80 and 110°C without the possibility of producing energy via the formation of protons or a sodium gradient, or without preventing the leakage of small hydrophobic molecules from the cell? The only possible conclusion is that they would contain special lipids making their membrane impermeable, even at high temperatures.

Biochemists who in the 1970s and 1980s were interested in the lipids of hyperthermophiles had a good life: they quickly reaped a harvest of astonishing discoveries. At first glance, however, the membranes of these microbes were not fundamentally distinguishable from those of mesophilic organisms: They are also made up of glycerolipids that form a hydrophilic film covering a hydrophobic core. All the same, their lipids are very special.

University students in the first year of biology learn that the glycerolipids of all living beings, from bacteria to humans, are esters of glycerol and fatty acids. This isn't exactly true. In the 1960s, Canadian scientists discovered another type of glycerolipids in *Halobacterium salinarium*, which lives in environments where the concentration of salt is very high.[57] If the glycerolipids of this halophile microbe are indeed formed of a molecule of glycerol and two tails, these tails are not formed by fatty acids, but by alcohols with long carbonaceous chains called isoprenol. Whereas the carbonaceous chains of fatty acids are linear, those of isoprenol are considered branched: every five atoms of carbon, one of the two hydrogen atoms is replaced by a methyl group (CH_3). Furthermore, the association of the alcohol functional groups of glycerol and of those of isoprenol is established through an ether bond, more stable at high temperatures than the ester bond of "common" lipids (figure 3.9).

How can this be useful to halophile microbes? They don't live at very high temperatures because they proliferate in saltwater lakes, salt marshes, or the Dead Sea at temperatures that never go above 40 to 50°C. The question remained for a time without an answer. Then, in 1972, when the American biochemist Thomas Langworthy undertook an analysis of the chemical composition of the membrane of *Thermoplasma acidophilum*, which had just been isolated by Brock (see chapter 1), it appeared that this thermoacidophile microbe possessed lipids identical to those of the halophile microbes of the Dead Sea.[58] These same lipids were found again two years later in *Sulfolobus acidocaldarius*, also isolated by Brock.[59] This time we understood the importance of lipids that did not risk "losing their head" to these two thermophile organisms, but that didn't answer the question raised by their presence in halophiles. The question was resolved five years later, when Woese showed that *H. salinarium*, *T. acidophilum*, and *S. acidocaldarius* were not bacteria, but archaea.

Woese immediately concluded that the atypical lipids discovered in these three organisms were simply typical of archaea. This was soon confirmed by Langworthy, who showed that the same lipids were present in methanogenic microbes (the first proven archaea), most of which live at low temperatures and in the absence of salt.[60] In other words, all archaea, whether they live in our intestines (methanogens) or in the Arctic, salt lakes, or hot springs, possess lipids formed by ether bonds between a glycerol molecule and two isoprene chains. As a result, the mesophilic archaea, like their hyperthermophilic siblings, possess membrane lipids with solid heads.

But lipid specialists had more microbe-from-hell surprises up their sleeves. In his 1974 paper, Langworthy showed that *Sulfolobus acidocaldarius* not only possessed ether lipids but also possessed giant "tetraether lipids" with two heads (bicephalic; figure 3.8b). In them, two "head-to-tail" glycerol molecules are connected by two long carbonaceous chains of isoprenol formed by the combination of forty carbon atoms (versus around twenty for "normal" lipids with one head). Therefore, in such lipids, the two lipid layers of the *Sul-*

folobus membrane are fused together to form a monolayer whose thickness is identical to that of a "standard" bilayer (figure 3.8). When biochemists were able to systematically analyze the membranes of new hyperthermophile archaea discovered in the 1980s to 1990s by Zillig and Stetter, they noticed that most of them possessed giant tetraether lipids capable of forming monolayer membranes. This strongly suggested that these atypical molecules played an essential role in an ability to adapt to very high temperatures. Indeed, it has been shown that the percentage of bicephalic lipids increases as the temperature rises in the methanogen from hell, *Methanocaldococcus jannashii*.[61] The Japanese lipid expert Yosuke Koga gave archaeal diether lipids the name "archaeol" ("the alcohols of the archaea") and archaeal tetraether lipids the name "caldarchaeols" ("the alcohols of the archaea of heat").[62] However, the name "caldarchaeol" should not be taken literally, since they have been found in many mesophilic and even some psychrophilic archaea, whereas they are missing in some hyperthermophiles such as *Methanopyrus kandleri*. It seems that all archaeal lipids are perfect for living at any temperature from 0 to 110°C.

Lipids with solid heads (ether bonds) have also been found in some bacteria (figure 3.9). However, they are formed by the association of glycerol and fatty alcohols and not of glycerol and isoprenol: there are thus typically bacterial lipids (they are not isoprenoids) that have ether bonds resistant to high temperatures. It is likely no coincidence that they are found in the membrane of the most hyperthermophilic bacteria, such as *Aquifex pyrophilus* and *Aquifex aeolicus*, that can divide in conditions up to 95°C. Like the solid heads, giant bicephalic lipids are perhaps not specific to archaea, either. In 1979 the English scientist Robert Klein isolated, in a bacterium of the genus *Butyrivibrio*, fatty acids that he christened "diabolical acids" (not in reference to hell, but because they were difficult to characterize).[63] These diabolical acids could theoretically form giant lipids with two heads of bacterial type. As if by chance, the diabolical acids were then rediscovered in certain hyperthermophile bacteria, in particular *Thermotoga maritima*, which seems to confirm their "infernal" nature.[64]

Why Won't Life Likely Go Beyond 110 to 120°C?

If bacteria can also "concoct" lipids that do not lose their heads at high temperatures, and even perhaps giant bicephalic lipids, why haven't they been able to compete with archaea between 95 and 113°C? The response to this question came out of the work of Konings's team in Groningen (Netherlands). These biochemists measured precisely the permeability to protons and ions of vesicles formed by the lipids of *Thermotoga maritima* (a hyperthermophilic bacterium) and that of vesicles formed from lipids of *Sulfolobus acidocaldarius* (a hyperthermophile archaea). The verdict was that the vesicles formed from lipids of *Sulfolobus* (which essentially contain caldarchaeol) tolerate very high temperatures much better than those formed from lipids of *Thermotoga maritima* (in spite of their lipids mimicking those of archaea). The lipidic vesicles of archaea type (*Sulfolobus*) remained perfectly impermeable to protons up to 100°C (figure 3.8b), whereas those of bacterial type (*Thermotoga*) were transformed into a sieve at that temperature.[65]

Why are archaeal membranes capable of remaining impermeable at infernal temperatures? It seems that at very high temperatures, protons or sodium ions can make a path through the classic membranes thanks to the disordered movements of the lipid tails, which create gaps through which they can slip. In a membrane formed of archaeal lipids (archaeol or cardarchaeaol), the methyl groups that line up all along the lipid tails in archaea (one every five carbons on the isoprene chain; see figure 3.9) should play a key role in preventing the uncontrolled passing of protons or sodium ions, forcing them to go toward the "pump" that awaits them to create ATP. Furthermore, in caldarchaeol, it is as if these tails are ligated: they can no longer move, which probably leaves fewer opportunities for the protons or sodium ions to slip through them. This is probably why, unlike bacteria, hyperthermophile archaea can continue to create ATP in large quantities at a temperature of 100°C. Let's note, however, that at 105°C things start to go wrong, and at 110°C even the vesicles formed by caldarchaeol begin to allow sodium ion protons to diffuse freely. Is it by chance if these temperatures correspond to the "hottest" beings of terrestrial life? I don't think so. I do think that the permeability of the membranes constitutes the true Achilles' heel of microbes from hell, and that we will never find microbes living at 130 or 150°C unless we imagine a type of lipid and membrane still completely unknown in the living world.

I can't end this discussion of membrane lipids without mentioning another crucial difference between the lipids of archaea, whether they are hyperthermophile or not, and those of bacteria and eukaryotes: their asymmetry. What does this mean? You perhaps know of the first experiments done by Pasteur, which led him to the discovery of two forms, right and left, of a molecule called tartaric acid. Just like that acid, many organic molecules can exist in two forms that are the mirror image of each other (like the right hand and the left hand). For this, one of their carbon atoms needs to be asymmetrical—that is, connected by its four bonds to four different groups of atoms. This situation is found in amino acids, which are all of a "left" shape in proteins, and in sugars, such as ribose, which are all of a "right" shape in nucleic acids. This

asymmetry of the living being is not mysterious: it is due to the spatial structure of the active site of enzymes or ribozymes, which can recognize only one
of the two shapes (like a glove can "recognize" only one of our two hands).
Thus the ribosome can form a peptide bond only between the "left" amino
acids. Glycerolipids are also asymmetrical, but unlike sugars and amino acids,
they do not have the same shape throughout the entire living world: they are
in the opposite orientation in archaea, compared to bacteria and eukaryotes
(figure 3.9). Indeed, the enzyme that creates the bonds between the glycerol
and the long carbonaceous chains in archaea is not related to the one that
carries out this same work in bacteria and eukaryotes.[66] They originated independently in the course of evolution. On this level, archaea thus form a group
that is completely separate from the rest of the living world. The discovery
of their very specific lipids has greatly contributed to having the concept of
"archaebacteria" accepted by a first group of biologists in the 1980s.

Why are archaeal lipids so different from bacterial and eukaryotic lipids?
What kind of lipid was present in LUCA and earlier, in the archaeal or bacterial type? What are the most ancient type of lipids? All these fascinating
questions have not yet found satisfactory answers. They are directly related to
the question that has caused the most ink to flow on the subject of microbes
from hell: that of their origin and of the origin of living beings in general. We
will approach these questions in the next chapter. We will begin with a little
detour, where we will get to know better a figure central to our story—Carl
Woese, the king of archaea—during his encounter with another monarch, the
king of Sweden.

The Universal Tree of Life: Where to Place Microbes from Hell and Their Viruses?

September 24, 2003: Stockholm—Carl Woese Is Awarded the Crafoord Prize

Under the crystal chandeliers of the Grand Hotel in Stockholm, the banquet is in full swing. In the middle of the room is the table of the king of Sweden and of Carl Woese, who this afternoon has received from the hands of His Majesty the Crafoord Prize (as prestigious as the Nobel Prize for those in the know, but completely unknown by the media)[1] (figure 4.1, left).

At the neighboring table, Mrs. Woese is talking to the queen. Earlier, we had waited for some time, standing next to our tables, for the monarchs to arrive. Were they stuck in a traffic jam? Anyway, one doesn't sit down before the king does. Now we're enjoying a fine meal in this mythical place, mythical because all post-Nobel and post-Crafoord receptions have taken place in this building of characteristic late nineteenth-century architecture. And now a lovely young woman in a long, flowing mauve gown stands. She moves to the podium in the center of the room, takes the microphone, and asks for silence: "Your majesties, members of the Royal family, members of the Royal Academy of Sciences, members of the Crafoord family, Professor Carl Woese. . . ."

This is the third time she has done this. The first was to propose a toast to the king and queen; the second was to announce a speech by the president of the Royal Swedish Academy of Sciences. What will she say to us this time? "We invite Professor Carl Woese to speak."

Good—the hero of the day is finally going to speak. He walks slowly to the podium. He's rather short, with a somewhat gloomy air about him, but his eyes are sparkling with mischief. He begins, "Your majesties, members of the Royal family, members of the Royal Swedish Academy of Sciences, members of the Crafoord family, Professor Carl Woese. . . ."

And he turns his head, looking for himself in the room. A moment of

FIGURE 4.1. Woese receiving the Crafoord medal from the king of Sweden in Stockholm in 2003 (*left*) and holding the film showing the oligonucleotide pattern of the 16S rRNA molecule from the first methanogen analyzed that turned out to be an "archaebacterium" (*right*).

doubt and incredulity ripples through the assembly. Is this American mocking the traditions of the old European monarchy? But then he smiles and begins again. A few people applaud. It's lucky he didn't receive the Nobel Prize, because the press would have been there.

I was fortunate, along with some other colleagues, to be invited to share this moment of consecration with Woese and to give two lectures, one in Lund and the other in Stockholm, on the occasion of these "Crafoord days." For my lectures I chose the title "Archaea: A Gold Mine for Molecular Biologists" because archaea were the source of a harvest of particularly exciting discoveries. For example, we will see in the next chapter that they have enabled our Orsay team to discover an unexpected evolutive link between archaea, sex, and the size of plants. Zillig's first observations on the resemblance of RNA polymerases in archaea and eukaryotes were followed by a great number of discoveries that showed how close the great molecular mechanisms in archaea and eukaryotes really are. These observations led to another question that we will look at in the final chapter: Are archaea our ancestors? All these discoveries would not have occurred so rapidly without Woese's visionary work.

Woese was a fascinating character, rather secretive, who tended to keep his distance from the scientific community. He almost never attended meetings, and he refused most invitations. He never recovered from the refusal of several very well-known Anglo-Saxon evolutionists to accept his great discovery: the living world is not divided into two domains, but into three. How could he, an ex-physicist turned molecular biologist, have had the audacity to challenge the dogma accepted at the time by so many biologists? The resistance was strong and came from many different sides. Some, especially cellular biologists, were adamant about conserving the eukaryote/prokaryote dichotomy. For them, dividing the world into three was a "molecularist" whim: what was most important wasn't the ribosome, but the nucleus (having one or not). It is difficult to get rid of old habits. Furthermore, many biologists remained skeptical: was it really possible to retrace the history of all living beings from a single molecule, the 16S ribosomal RNA, when the genome of the smallest bacteria already includes more than five hundred genes? All of this probably explains why Woese had to wait twenty-six long years (from 1977 to 2003) to be knighted by the king of Sweden and receive the Crafoord Prize.

If the Swedish Academy decided to award the Crafoord Prize to Woese, it's because the facts ultimately proved him right. His ingenious intuition was ultimately validated by the "genome revolution" that began in the mid-1990s. As I briefly mentioned in chapter 2, the analysis of proteomes (all the proteins coded by genomes) showed there are indeed three versions for each of the universal proteins: one per domain. Some authors have noted that the universal proteins still represent only a small percentage of the total number of proteins present in the proteome of an organism, no more than 1%, and that for the other proteins, often "bacterial" versions were found in an archaea, or vice versa.[2] This phenomenon is the result of the transfers of genes, discussed in chapter 2. All the same, we will see that this did not prevent the evolutionists from reconstituting the history of organisms along these great lines. The results obtained supported the work of Woese and his colleagues, which was nonetheless based on only a single gene, or less than 0.1% of all the genes of an organism.

Upon arriving in Lund, Sweden, for the first of three "Crafoord days" that preceded the award ceremony, I was a bit concerned. I had never had the opportunity to talk with Woese directly and I didn't really know what he was going to think of my presence there. The evening of my arrival I met the microbiologist Karl Stetter, the father of hyperthermophiles, at the hotel bar, and I mentioned my concern to him: what Woese knew of my ideas on evolution might not please him very much. In several publications I had argued against the idea of a hot origin of life and of a direct link between that origin and current hyperthermophiles, whereas Woese was a hot partisan of that hypothesis, as was Stetter. "Just a little different," repeated Stetter with a loud laugh, passing me a glass of beer. We had at least had a frank discussion on the subject a few years earlier, and we later became good friends.

My encounter with Woese ultimately went very well. In my lecture I had shown a slide that displayed the universal tree he had used to illustrate his popular article on archaebacteria published in *Pour la Science* in 1982,[3] which was when I discovered his work (see figure 1.6). Woese was, it seems, very touched that someone would still remember that popular article. The following year, he wrote a very kind letter in support of my candidacy to the Institut Universitaire de France, a virtual institution that allows select professors to reduce their teaching loads so they have more time to devote to research. I think his letter must have helped me win this nomination, which was very important to me. The teaching load of professors in France is particularly heavy, which is a great handicap for scientific research in our country. Without my position at the Institut Universitaire de France, I would probably never have succeeded ten years later in obtaining the European funding that today enables me to pursue my research at the Institut Pasteur. Thank you, Carl Woese.

When I met Woese, I realized that he liked to talk with scientists who were passionate about evolution even if they didn't agree with him. Of course, we retained our positions concerning the origins—hot or cold—of life. In any case, whatever I might have said, he liked the hypothesis of a hot origin, and he wasn't alone: the idea of an original hell had succeeded that of a lost paradise. Also, as we will see in detail a bit later, this idea seemed validated by the universal tree of life reconstructed, at the beginning of the 1990s, through the work of Woese and his team. This tree was characterized by the almost exclusive presence of microbes from hell at the level of its roots, its trunk, and its lowest branches. Woese's process, which was very rigorous, seemed to demonstrate scientifically that LUCA, the last common ancestor of all current living beings, had lived in hell. Most of the participants at the Stockholm ceremony were thrilled by this conclusion, which was certainly seductive but proved to be probably wrong. We will see that LUCA was likely not a hyperthermophile, which, in my opinion, doesn't detract at all from its importance. Indeed, I will present a new hypothesis suggesting that adaptation to infernal temperatures might have given birth to two of the three domains: bacteria and archaea. Before seeing how I arrived at this conclusion, let's return to the story of the discovery of archaea, that major twentieth-century discovery in biology, which earned Carl Woese the right to walk under the crystal chandeliers of the Grand Hotel in Stockholm.

A Rosetta Stone for Biologists: The 16S Ribosomal RNA

We saw in chapter 1 that Woese had chosen as a molecule, a witness to evolution, one of the RNA molecules present in ribosomes, those cellular factories that manufacture our proteins (figure 4.2).

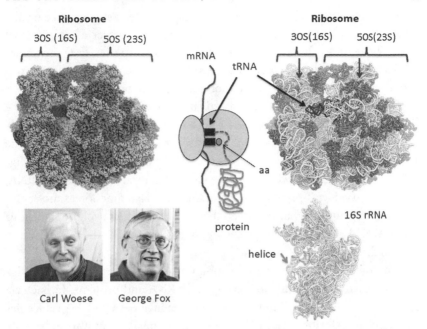

FIGURE 4.2. Ribosomal RNA: The Rosetta stone of cellular evolution. Two models of the bacterium *Thermus aquaticus* ribosome obtained from crystallographic analysis. In the left model, all atoms are in light gray for ribosomal RNA, dark gray for proteins, and black for transfer RNA (tRNA). In the right model, the polypeptide chains are in dark gray, the RNA backbone is in light gray, and tRNA are in black. You can recognize the double helical structure of many RNA regions. The active site is hidden within the large subunit and surrounded only by RNA. The structure of the 16S ribosomal RNA (bottom right) has been extracted from the above model. The middle sketch illustrates the process of translation. The growing polypeptide chain escapes from the ribosome large subunit via a tunnel (*dotted line*).

Why that choice? First, because of the universality of ribosomal RNA: remember that all cellular organisms possess ribosomes to synthesize their proteins. Next, for the universality of their functions: within ribosomes, ribosomal RNA intervenes on several levels in the synthesis of proteins, and the different functions ribosomes perform have been preserved throughout evolution. As we saw in the previous chapter, these functions are determined by the structure in space of ribosomal RNA, which itself depends on their nucleotide sequences. These latter, very long series of nucleotides A, U, G, or C (around 1,500) have thus been well preserved during evolution. All the same, this preservation is not absolute: many portions of the sequence of ribosomal RNA can vary, following mutations, without affecting their function. In other words, when a mutation occurs in one of these regions of the molecule, the mutant survives, it has descendants, and the mutation is then "fixed." Such mutations are called neutral, and at first approximation, one might think they accumulate with a constant speed during evolution: this seems logical, since they are produced by chance and confer no selective advantage or disadvantage to the individual that carries them. When two species diverge (that is, after being individualized from an initial ancestral species, they each independently pursue their evolution), the sequences of their ribosomal RNA will also diverge, through the accumulation of neutral mutations.

Thus the more time that has passed since the divergence, the greater the differences in the sequences of their ribosomal RNA. These RNA can be considered true molecular clocks and tentatively be used to build phylogenetic trees retracing the history of species containing these RNA molecules (figure 4.3). This reasoning, which we have just shown regarding ribosomal RNA, can also be used with proteins. Some of them can also function as molecular clocks if their function has remained the same throughout evolution; the four letters AUGC in the RNA sequences are simply replaced by the twenty amino acids in the sequences of proteins.

Ribosomes are formed by two subunits, one small and one large (figure 4.2). Each corresponds to a complex of RNA and proteins. In bacteria and archaea, the large subunit of the ribosome includes two RNA molecules called 23S and 5S ribosomal RNA, whereas the small subunit includes only one ribosomal RNA, called 16S ribosomal RNA. In scientific jargon, the letter S refers to the name of a Swedish scientist, Theodor Svedberg, who was the first to use a high-speed centrifuge to separate macromolecules by subjecting them to a gravitational field. This is the method used to separate the different ribosome subunits or the different ribosomal RNA molecules. The RNA are placed on the surface of a concentration gradient formed by a sucrose solution whose concentration increases from the surface to the bottom of the centrifuge tubes. Centrifugation is then stopped before the RNA falls to the bottom of the tube. The RNA will have more or less fallen in the gradient according to their shapes (aerodynamics) and sizes, which can be quantified by attributing each with a "coefficient of sedimentation" expressed in Svedberg. The higher the S value of a macromolecule, the faster it falls during centrifugation. Ribosomal RNA molecules having almost all the same shape, the 5S RNA molecule is thus smaller than that of the 16S ribosomal RNA, which is itself smaller than that of 23S ribosomal RNA. The RNA of eukaryotic ribosomes, which are bigger than those of bacteria or archaea, "measure" respectively 5S, 18S, and 28S. In eukaryotes, the large subunit of the ribosome includes an additional RNA molecule, RNA 5.8S, which is absent in archaea and bacteria.

When Woese began his research program, it was difficult to envision the study of the three ribosomal RNA molecules at the same time. He chose 16S ribosomal RNA because it contains more sequential information than

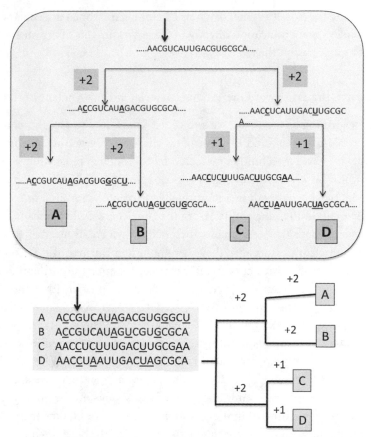

FIGURE 4.3. Evolution of ribosomal RNA sequences. The numbers of mutations (*underlined*) at each step are indicated in gray squares. Scientists can only access sequences of organisms living today (A, B, C, D). They can align these sequences by putting on the same column (*arrow*) homologous nucleotides that are derived from the same ancestor. From this alignment, they can calculate the number of mutations that occurred between two sequences and reconstitute a phylogenetic tree that recapitulates the evolution of the organisms. This is an oversimplified example. The same mutation can (i) occur independently in different lineages, (ii) accumulate more rapidly in some lineages, and (iii) reverse (from A to C, then C to A), making phylogenetic reconstruction sometimes very tricky.

RNA 5S, while also being easier to analyze than RNA 23S. His decision has considerable historical significance: 16S ribosomal RNA became the Rosetta stone for biologists, enabling them to decode the history of living beings, even if, as we will discover a bit later, one must not always have blind faith in it. Even today, when you discover a new microbe, the sequencing of its 16S ribosomal RNA is a required passage to be able to publish its description, because that enables it to be situated in relation to those that are already known (even if it is increasingly easy to obtain the complete sequence of the genome of the microorganism in question). We will see that 16S ribosomal RNA has also become the Rosetta stone for a new discipline, microbial molecular ecology, by enabling identification directly in the environments of organisms that cannot be cultured in the lab.

Return to the Trinity of Life: "Archaebacteria," a Very Ancient Lineage?

At the beginning of the 1970s, we still didn't know how to rapidly sequence RNA or DNA.[4] Woese and his colleagues succeeded in getting around this difficulty by using a technique perfected by the English biochemist Frederick Sanger. This involved creating a "genetic imprint" of 16S ribosomal RNA molecules by cutting them in a reproducible way into a hundred or so small fragments called oligonucleotides, from the Greek *oligos*, which means "small in number." These RNA fragments included only a small number of A, U, G, and C nucleotides because they were obtained through the action of a ribonuclease that always cuts the RNA molecule after the nucleotide corresponding to the letter G. For example, the portion of an RNA molecule corresponding to the sequence AUUUCGUUGAAU would have given the three oligonucleotides AUUUCG, UUG, and AAU. The genetic imprint of an organism corresponds to all the oligonucleotides identified in it. These oligonucleotides, previously made radioactive (see box 4.1), were separated from each other through chromatography depending on their size and composition. They could be seen in the form of black spots on the photographic film held by Woese himself in the picture in figure 4.1, right. One then had to compare the imprint of the species being studied with those of all the other species already analyzed. This method, clever but rather complicated, demanded an enormous amount of meticulous work (see box 4.1). It enabled the quantification of the evolutive divergence between two species, A and B, through the calculation of a coefficient of similarity (S_{AB}) whose value translated the resemblance of their catalogs of 16S ribosomal RNA: the S_{AB} varied thus between 0.1 for two very distant species, and 1 for two identical species.

When George Fox, a postdoctoral researcher on Woese's team (figure 4.2), analyzed the 16S ribosomal RNA of a methanogenic "bacterium," *Methanobacterium thermoautotrophicum* ΔH, in May 1976, he noticed that its genetic imprint was quite different from those of all the bacteria they had studied up to then. These differences were visible to the naked eye when one compared the arrangement of black spots corresponding to the oligonucleotides on the photographic film (see box 4.1). Many of the spots observed until then in all bacteria were absent, whereas new spots that had never been seen before had appeared. For a time, Fox even wondered if he had sequenced the right RNA, or if he had separated the 16S and 23S ribosomal RNA before digestion by the

BOX 4.1. THE S$_{AB}$ METHOD

To compare the 16S ribosomal RNA molecules of different organisms without having to sequence the DNA, Woese and his team first had to cultivate a large quantity of organisms to isolate their 16S ribosomal RNA. These RNA molecules were then treated with an enzyme, called ribonuclease T1, systematically cutting the sequence of the RNA to the right of a G. The different oligonucleotides present in the obtained mixture (a few dozen) were then separated on the surface of a sheet of paper according to their chemical properties using a method called two-dimensional chromatography. The cells out of which the 16S ribosomal RNA molecules emerged had been cultured in a medium containing a radioactive isotope of phosphorus. The oligonucleotides had thus incorporated this marked element and had also become radioactive. They then appeared in the form of black spots on a photographic film placed for a time above the sheet of paper. The arrangement of these spots was characteristic of the species whose 16S ribosomal RNA was being studied. The oligonucleotides corresponding to each spot could then be eluted (removed) from the sheet of paper and their sequence identified by a combination of chemical and enzymatic methods. Many small oligonucleotides (AG, CCG, UAAG, etc.) were found by chance in the catalogue of all species and were thus not interesting for the analysis. Woese therefore chose to study only oligonucleotides with a length of at least six nucleotides (AUUUCG in the preceding example). In these conditions, if the same oligonucleotide was found in two different species, it was probably not there as the result of chance, but rather because they had inherited it from a common ancestor. As a consequence, the closer two species were in kinship—that is, the more recent their divergence from this common ancestor—the greater the number of large oligonucleotides they had in common. To quantify the resemblance between two organisms, A and B, Fox calculated a coefficient of similarity, the famous S$_{AB}$. It corresponded to the number of large oligonucleotides common to two species (N$_{AB}$) multiplied by two and divided by the total sum of the large oligonucleotides present in the catalogue of each of the two species, N$_A$ and N$_B$, or S$_{AB}$ = 2 N$_{AB}$ / (N$_A$ + N$_B$). We see that for two identical species, N$_A$ = N$_B$ = N$_{AB}$, the S$_{AB}$ is equal to 1, whereas the farther apart two species are, the more it will tend toward 0.

ribonuclease. But when he repeated the experiment over three other species of methanogens in 1976, he obtained the same results. The spots absent in the first methanogen studied were still absent in the three others; the new spots that appeared were common to the four methanogens. This was the "eureka" moment.

Four years after Woese won the Crafoord Prize, I had the opportunity to

	E1	E2	E3	B1	B2	B3	B4	A1	A2	A3	A4	A5	A6
E1	1	0.29	0.33	0.05	0.08	0.11	0.08	0.11	0.11	0.08	0.10	0.07	0.04
E2	0.29	1	0.36	0.10	00.06	0.09	0.11	0.11	0.08	0.08	0.09	0.07	0.09
E3	0.33	0.36	1	0.06	0.07	0.06	0.10	0.1	0.01	0.09	0.11	0.08	0.07
B1	0.05	0.10	0.06	1	0.26	0.28	0.21	0.11	0.12	0.07	0.07	0.07	0.09
B2	0.08	0.06	0.07	0.25	1	0.26	0.20	0.11	0.13	0.06	0.10	0.07	0.09
B3	0.11	0.9	0.9	0.25	0.26	1	0.31				0.10	0.10	0.10
B4	0.08	0.11	0.06	0.21	0.20	0.31	1	0.14	0.12	0.10	0.12	0.06	0.07
A1	0.11	0.10	0.10	0.11	0.11	0.11	0.14	1	0.51	0.25	0.34	0.17	0.19
A2	0.08	0.13	0.09	0.07	0.06	0.10	0.10	0.25	1	0.32	0.31	0.13	0.21
A3	0.08	0.07	0.07	0.12	0.12	0.10	0.12	0.30	0.24	1	0.29	0.16	0.23
A4	0.10	0.09	0.11	0.07	0.10	0.13	0.12	0.34	0.29	0.28	1	0.19	0.23
A5	0.07	0.07	0.06	0.07	0.07	0.10	0.06	0.17	0.13	0.16	0.19	1	0.13
A6	0.08	0.09	0.07	0.09	0.05	0.10	0.07	0.19	0.21	0.23	0.23	0.13	1

E1 : Yeast (Saccharomyces cerevisiae)
E2 : Plant (Lemna minor)
E3 : Homo sapiens
B1 : Escherichia coli
B2 : Bacillus firmus
B3 : Cyanobacteria (Aphanocapsa)
B4 : Chloroplast (Lemna minor)
A1 : Methanobacterium thermoautotrophicum
A2 : Methanogen sp./
A3 : Methanosarcina barkeri
A4 : Halobacterium halobium
A5 : Sulfolobus acidocaldarius
A6 : Thermoplasma acidophilum

FIGURE 4.4. Selection of S_{AB} values obtained by Woese in comparing six archaeal species with four bacterial and three eukaryotic species (adapted from Woese and Fox, 1997). The S_{AB} values for all couples are indicated. The inserted picture shows Woese's thumb pointing to an oligonucleotide specific for archaea, due to a specific base modification in one of the nucleotides.

see him again during a colloquium in 2007 organized at the University of Illinois at Urbana, where he worked, for the thirtieth anniversary of the discovery of archaea. I remember my emotion when he pulled out of its box the historical film corresponding to the imprint of *Methanobacterium thermoautotrophicum* ΔH. This film (figure 4.1, right) was almost as tall as I am. Woese pointed at one spot in particular: "You can see this spot, Patrick. Its location tells us that something special was going on. It should not have been there; we never saw it before in bacterial catalogues. It is because one of its bases contains a chemical modification unique to archaea" (figure 4.4).

This first visual impression, which revealed the strangeness of methanogens, was confirmed by the S_{AB} calculation (see the table in figure 4.4). As a general rule, the S_{AB} of two distantly related bacteria was between 0.2 and 0.3. But in the couple formed by one of the four species of methanogen "bacteria" studied and any other bacterium, the S_{AB} value was always 0.1 (between 0.06 and 0.14). To Woese and Fox's great surprise, methanogens, although prokaryotes, thus showed no detectable link of kinship with bacteria. Later they would complete this work by obtaining (with even more difficulty) the imprint of several eukaryotic 18S ribosomal RNA (the homologues of bacterial 16S ribosomal RNA). Here, too, whereas the S_{AB} value for the human-plant or human-beer yeast couples was 0.33, those values were always around or lower than 0.1 for couples including a bacterium and a eukaryote or a methanogen and a eukaryote.

These results thus revealed the existence of three types of 16S ribosomal RNA/18S, and thus three types of ribosomes in the living world: Woese had shown that this world was not divided into two domains (prokaryotes and eukaryotes), but into three. He had revealed the existence of a new group of living beings on our planet. This work had been funded in part by NASA within the framework of a program seeking life on Mars. Woese had probably justified his participation in this project by announcing the perfection of a universal method of detection and classification of living beings (although there is no chance that potential Martians have 16S ribosomal RNA). In any event, although Woese didn't discover life on Mars, he did demonstrate a new form of life on Earth—not too bad.

At the time, Woese and Fox made another remarkable observation: the S_{AB} values calculated for some couples of methanogens were very weak: 0.3. In other words, two methanogen organisms could be as distanced from each other on the evolutive level as were the most divergent of eukaryotes (humans and plants, for example), or bacteria (cyanobacteria and *Escherichia coli*). When the S_{AB} method led to the regrouping of some halophile and thermoacidophile microbes (*Sulfolobus acidocaldarius* and *Thermoplasma acidophilum*; see chapter 1) with methanogens, they once again noted that large evolutive distances could separate members of the new domain of the living that they had just discovered (figure 4.4). Thus the S_{AB} values of the couples formed by *S. acidocaldarius* and any other archaebacteria were between 0.13 and 0.19. This observation in part explained why the name "archaebacteria"—"old bacteria"—was chosen at that time to name these organisms "of the third type" (in addition to the metabolism of methanogens adapted to the primitive Earth atmosphere; see chapter 1). Indeed, the weak S_{AB} value observed between *S. acidocaldarius* and other archaea suggested that the lineages leading to these two organisms had been individualized well before those leading to bacteria or eukaryotes (the S_{AB} values of two bacteria or two eukaryotes, even the most distant among them, always being above 0.2).[5]

From 16S Ribosomal RNA to 16S DNA Genes: Always More Sequences

The historic S_{AB} method was quickly abandoned when in the 1980s a sequencing method for the DNA molecule was perfected, again by Sanger. It was this method that would enable Woese's team to construct the first universal tree of life. Let's note that twenty years earlier, Sanger had invented the technique that enabled the order of amino acids in the sequence of proteins to be determined. This exceptional scientist was awarded a Nobel Prize for each of these sequencing methods (in 1958 and in 1980). We will see, moreover, that the method Sanger perfected for DNA sequencing is at the foundation of the genome revolution. For all evolutionists, Sanger's method had an immediate impact. It enabled the direct sequencing of the gene coding for 16S ribosomal RNA without having to go through laborious steps of purification and radioactive marking of ribosomes. Just like the messenger RNA (see chapter 3), ribosomal RNA are transcribed from a gene onto the DNA of a chromosome. The sequence of this gene is thus identical to that of ribosomal RNA. From a practical point of view, the sequencing of the gene coding for 16S ribosomal RNA is much simpler and faster to implement than the S_{AB} method.

The use of the method of DNA amplification through polymerase chain reaction (PCR; see the prologue) at the end of the 1980s made the exercise even easier: one could directly amplify, at high temperature, the gene coding for 16S ribosomal RNA of an organism from a small sample of its DNA by using the DNA polymerase extracted from one of our infernal microbes (the bacterium *Thermus aquaticus* or the archaea *Pyrococcus furiosus*), then sequence it. Today we have complete sequences of several thousands of 16S ribosomal RNA molecules, which can be compared using different computer programs enabling the measurement of the evolutionary distances among species. To achieve this feat, the sequences are aligned beneath each other to identify the positions that descend from the same ancestral nucleotide in the 16S ribosomal RNA molecule that existed in the common ancestor of all organisms being studied (figure 4.3). Depending on the organisms, these positions will be occupied by one of the four nucleotides A, U, G, or C. From that data, we can calculate the percentage of identity between two sequences (that is, the percentage of positions that are occupied by the same nucleotide, the same letter), which gives us an indication of their resemblance. By comparing all the sequences two by two, the computer (or rather the program in it) constructs a tree in which the length of the branches connecting different species is directly proportional to the global resemblance of their 16S ribosomal RNA (figure 4.3). The same approach enables the construction of trees based on protein sequences, the positions analyzed corresponding, in this case, to amino acids. In both cases, specialists will tell you that my presentation of the construction methods for trees is a bit simplistic (to say the least). That's true; many programs enable computers to construct trees without relying uniquely on the global resemblance of sequences, but, for example, on the most probable paths enabling the passage from one sequence to another while minimizing the number of mutations produced in the entirety of the

tree (this is called parsimony).[6] For the moment, let's remember that the sequences contain a message from the past, which allows scientists to hope to
reconstruct the history of the living world more or less precisely.

When Microbiologists Become Ecologists

Before turning to the universal tree of life obtained by comparing 16S ribosomal RNA molecules, let's return for a moment to Stockholm. Among those
invited to the Crafoord days in 2003, we must mention Norman Pace. This
former student of Woese's belonged to the generation of "heroes" that perfected the methods for analyzing ribosomal RNA through genetic imprint-
ing. Pace was invited to speak about "microbial molecular ecology" at the
symposium organized in his mentor's honor. Pace is in fact one of the creators
of this new scientific discipline. After leaving Woese's laboratory, he had the
idea of using the 16S ribosomal RNA molecule as a marker to directly detect
microbes in the environment without having to culture them. This approach
became possible thanks to the methodological revolution introduced by the
direct sequencing of DNA and its amplification through PCR. Its principle
is simple: isolate the DNA present in the sample from the environment to
be analyzed (the water from a hydrothermal spring or from a purification
station, the humus from a forest, etc.), then specifically amplify all the genes'
coding for 16S ribosomal RNA using a thermostable polymerase DNA and
the PCR technique.

To amplify DNA, it is necessary to initiate the polymerization reaction
with a small primer of complementary DNA from a part of the gene that one
wants to amplify. This primer will be hybridized with the DNA to be amplified, enabling the polymerase to recognize the gene on which it must work.
By carefully choosing the right primer, one that can hybridize only with archaeal 16S ribosomal RNA or, on the contrary, one that can hybridize only
with bacterial 16S ribosomal RNA, one can specifically amplify either archaeal
or bacterial 16S ribosomal RNA molecules directly from the environment.
This approach has enabled the discovery of new microbes belonging to these
two domains, even before we are able to isolate them and cultivate them in
the lab. The teams of Pace and another American scientist, Edward DeLong,
were thus able to show that all the biotopes of the planet are inhabited by a
multitude of bacteria and archaea that up to then had gone unnoticed.[7]

We'd known for a long time that the observation of samples of water or soil
under the microscope revealed the presence of a greater number of bacteria
than we could culture in the lab from the same sample; this is an anomaly
in the numbering of bacteria. But we hadn't imagined the great number of

unknown microbes. The microbes we know represented probably only 0.1 to 1% of those existing around us. Suddenly, new groups of bacteria and archaea whose existence no one had ever guessed before were revealed. In most cases, these groups corresponded to organisms that we could not culture in the laboratory; we knew only the sequence of their 16S ribosomal RNA, which corresponded to their "phylotypes." Recall that the phenotype corresponds to the apparent characteristics of an organism and the genotype to the whole of these genes. The phenotype is determined by the interaction between the genotype and the environment. The phylotype designates the evolutive position of the organism. The analysis of the 16S ribosomal RNA sequences present in the environment enables the positioning of organisms having these sequences in phylogenetic trees constructed from the 16S ribosomal RNA of known organisms. It becomes possible to classify the organisms having these "environmental" sequences within a group of microbes already identified or, on the contrary, to consider them as representative of new groups even if we don't know anything else about them, even if we have never seen them under the microscope.

In the 1990s the total number of microbial groups exploded in a spectacular way thanks to the work of molecular ecologists. For example, in 2001 DeLong's team showed that around 30% of the biomass present in the oceans was made up of unknown mesophilic archaea whose 16S ribosomal RNA was related to that of some thermophile archaea.[8] This stunning observation set off a new hunt for archaea that ended in the discovery in 2005 of the first oceanic mesophilic archaea, *Nitrosopumilus maritimus*.[9] This very small archaea (*pumilus* means "dwarf" in Latin) was discovered by David Stahl's team in the seahorse tank in the Seattle Aquarium. We will see in the next chapter that these archaea correspond to a third large phylum of archaea that were only recognized in 2008. Many of these archaea oxidize ammonium and are widespread, not only in marine and freshwater biotopes, but also in most soils. Recently, other scientists have shown that these archaea are present on our skin.[10] Here we are far from the microbes from hell, but the existence and the diversity of these "cold" archaea shows that the third domain of the living world should not be considered a curiosity of nature, but represents an essential part of the biosphere, which until then had been hidden from us, "hidden before our eyes," to quote an expression used during the 2007 colloquium celebrating the thirtieth anniversary of the discovery of archaea.[11]

Today, the possibility of sequencing DNA on a large scale has gradually led molecular ecologists to abandon the sequencing of genes coding for 16S DNA for the massive sequencing of all DNA present in a sample coming from the environment.[12] This major technical advance has enabled the emergence of a new discipline that is currently very popular: metagenomics. All the same, here too the presence of a gene coding for 16S DNA (or 23S) in a fragment of DNA is the surest means for determining to which type of organism the fragment belongs, by correlating the genotype (all of the genome) to the phylotype (16S ribosomal RNA). Of course, the final objective is always to be able to culture new organisms detected by these molecular approaches to know what they resemble and to have access to their phenotypes.

You probably understand even better now why 16S ribosomal RNA can be considered the Rosetta stone of microbiology. If I had the space, I could say a lot more about the role that this molecule has played in the development of medical diagnostics for pathogenic bacteria. The possibility of identifying, through their genetic imprints, bacteria responsible for our illnesses, without having to wait for their culture in the lab (some are not able to be cultured, in any case) would save many lives. This is why Woese could just as easily have received the Nobel Prize in medicine as the Crafoord Prize. You might say that the difference between the two prizes is not very great. Like the Nobel, the Crafoord Prize is awarded by the Royal Swedish Academy of Sciences and the medal is bestowed by the king of Sweden. Like the Nobel, the reception takes place in the Grand Hotel in Stockholm and the check (rather generous) awarded to the recipient is about the same amount. Yes, but there is still a very big difference. Journalists do not attend the awarding of the Crafoord Prize, as they do the Nobel Prize, and the media never talk about it. So another opportunity to tell the world about the third domain of the living was missed.

Trees and Roots

Let's return to what had first motivated Woese to undertake his monumental work, to decode the history of the living world. Gary Olsen, one of Woese's students, in 1985 constructed the first universal tree by using all the complete sequences of 16S ribosomal RNA available at the time (before the introduction of PCR at the end of the eighties).[13] Let's look at this tree (figure 4.5a). We find the living world divided into three domains: bacteria, archaea, and eukaryotes. Let's also note that animals, fungi (the yeast *Saccharomyces cerevisiae*), and plants (Zea mays)—that is, practically the entire biosphere visible to the naked eye—represent only a very small part of the diversity of the living world. In this tree, the evolutionary distance between a human and a mushroom appears tiny compared to what sometimes separates two bacteria or two archaea. This shows that our view of the reality that surrounds us is biased: microbes form the very great majority of biodiversity. And yet this tree doesn't represent viruses, which don't have ribosomes and thus no

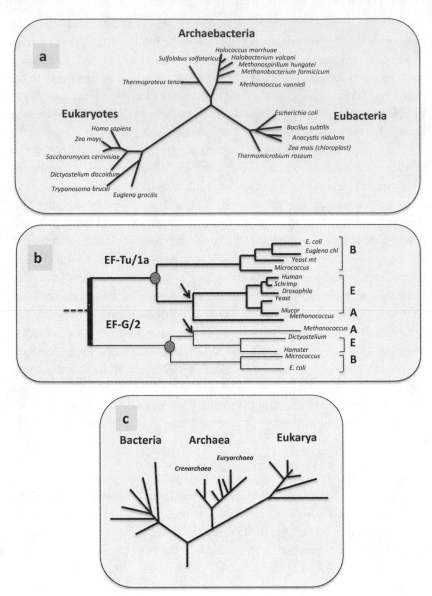

FIGURE 4.5. Historical phylogenetic trees. *A*, the first published universal tree based on 16S ribosomal RNA sequences comparison (Olsen and Woese, 1985). *B*, the two combined universal trees of the elongation factors in which the roots of the two trees (*gray circles*) are located between bacteria (B) and the last common ancestor of archaea (A) and eukarya (E) (Iwabe et al., 1989). The arrows indicate the points of trifurcation in both unrooted trees. *C*, the schematic tree published by Woese and colleagues to propose their novel nomenclature ("archaea" instead of "archaebacteria") (Woese et al., 1990). This tree combined the features of trees in *A* (the three domains based on rRNA phylogeny) and *B* (the root based on the elongation factors phylogeny).

ribosomal RNA. In fact, we will see at the end of this chapter that our tree is completely infected with viruses, from its roots to the thinnest of its branches.

You might object that the tree in figure 4.5a doesn't really look like a tree; rather, it looks like a sort of bush with three groups of foliage. In fact, it is a tree without roots. There is a reason for this: to place roots on a phylogenetic tree, you must know the direction of the flow of time. Evolutionists will say that to give roots to a tree there must be an external group, a group that is more distantly related to the groups one is studying than those groups are among themselves. For example, if one wishes to study the kinship relationships between humans, the gorilla, and the chimpanzee, one would take as an exterior group a monkey from the Americas, for instance, because we know that monkeys of the new world diverged from African monkeys well before the divergence between humans, gorillas, and chimpanzees. The roots of a phylogenetic tree are very different from the roots of trees of our forest, which can branch out infinitely and sink down to great depths beneath the trunk. In phylogenetic trees, the root corresponds to the deepest point of divergence corresponding to the separation of the tree into the first two branches. For example, in the case of the aforementioned monkeys, the root would be the point of divergence between the lineage that gave birth to the gorilla and that which gave birth to humans and chimpanzees (and yes, chimpanzees are our brothers and gorillas our cousins).

On reflection, one notices that it is not easy to situate the root of the universal tree of life, which is supposed to group all known species possessing ribosomes. How can we find an external group to enroot the universal tree? A Martian microbe, perhaps? No, there is little chance that the Martian microbe, if it exists, would be related to us in any way: it certainly does not possess 16S ribosomal RNA (unless there has been a cross-contamination between Mars and our planet that occurred through the exchange of meteorites at the very beginning of the history of the solar system, four billion years ago, which is a very improbable scenario). And an exterior group must share characteristics with the groups on the tree that we are seeking to enroot: it is necessary to make comparisons.

So how should we go about enrooting the universal tree of life? In 1989, the problem was resolved brilliantly by a Japanese scientist, Naoyuki Iwabe, on Takashi Miyata's team at Kyushu University. To root the universal tree, Iwabe and his colleagues used the sequences of two pairs of proteins present in all living beings: the two elongation factors (EF) involved in protein synthesis EF1 and EF2,[14] and the two catalytic subunits of the ATP synthases, the enzymes that produce the ATP at the level of the membranes (see chapter 2). In both cases, the two proteins of the pair, for example, EF1 and EF2, are the result of the duplication of a single ancestral gene that existed before LUCA. These two proteins are thus related and, by comparing their sequence in amino acids from one organism to another, we obtain two universal trees of life that are connected on the level of the event of duplication—pre-LUCA: one for EF1, the other for EF2. The tree constructed with EF1 can thus serve to root the tree constructed with EF2, and vice versa. In both cases, the results Iwabe obtained were identical. The root of each universal tree, corresponding to the position of LUCA, separated with a branch going toward bacteria and a branch going toward an ancestor common to archaea and eukaryotes (figure 4.5b). In this tree, archaea are thus our sisters and bacteria our cousins.[15] Iwabe obtained the same rooting (often called the bacterial rooting) with the catalytic subunits of the ATP synthases.[16] The latter result was confirmed independently that same year by Peter Gogarten, a German scientist working at the time in Santa Cruz, California.[17]

The results obtained by Iwabe and Gogarten were published in the journal *Proceedings of the National Academy of Sciences* in 1989 and immediately caused a great stir in the scientific community. In particular, the root obtained was immediately adopted by Woese, because it separated the archaea from bacteria. He saw there the opportunity to be finished once and for all with the confusion between the two groups of prokaryotes. No, archaea aren't slightly bizarre bacteria, but representatives of a completely different domain. Woese thus drew a new tree by combining the tree obtained by Olsen with the 16S ribosomal RNA and the "bacterial" root obtained by Iwabe and Gogarten from elongation factors and ATP synthases. This new tree (figure 4.5c), often called Woese's tree, today illustrates most studies published on archaea and student textbooks. Woese's tree was published in 1990 in *Proceedings of the National Academy of Sciences* jointly with Otto Kandler and Mark Wheelis. In that article, these authors proposed to abandon the term "archaebacteria" and replace it with "archaea." They also proposed replacing the term "primary kingdoms" with "domains" to designate the three major cellular lineages, demonstrated through the analysis of 16S ribosomal RNA: archaea, bacteria, and eukarya.[18]

In Woese's tree, archaea were divided into two large phyla based on the divergence of their 16S ribosomal RNA: Euryarchaeota and Crenarchaeota (figure 4.5c). Euryarchaeota (from the Greek *euryos*, "diversity") group archaea with phenotypes that are indeed very diverse: halophiles, methanogens, thermoacidophiles of the genus *Thermoplasma*, and hyperthermophiles of the genus *Thermococcales*, whereas the Crenarchaeota (from the Greek *crenos*, "springs as the origin of a river") included only thermophiles or hyperthermophiles, such as *Pyrolobus, Pyrodictium,* and *Sulfolobus.* The name Crenarchaeota, chosen by Woese, explicitly referred to the hypothesis of a hot origin of life, since this phylum was characterized by the "hot" phenotypes of its members, supposed to be an ancestral character at the origin of archaea. Notably, this division of archaea between these two major phyla had already appeared at the time of the S_{AB}. Remember, the S_{AB} of the pair formed by *Sulfolobus acidocaldarius* and any other archaea were very weak, around 0.15, which already clearly showed their great divergence (figure 4.4).

Did Life First Appear in Hell?

If there was a scientist who was happy with the positioning of the root of the universal tree of life proposed at the end of the 1980s, it was Karl Stetter. To understand why, let's examine a new tree, that of figure 4.6.

Stetter redrew it in 1995 from Woese's tree to emphasize the position of hyperthermophiles, his favorites. And so he drew in bold all the branches leading to a microbe from hell.[19] As we can see in figure 4.6, and it isn't surprising, most of them are located in the domain of the archaea and they form all basal branches of this domain (all emerging directly from the last common archaeal ancestor). Two lineages leading to hyperthermophilic microbes, *Thermotoga maritima* and *Aquifex pyrophilus,* also appear as basal branches in the bacterial part of the tree. Note that Stetter has drawn the branches leading to hyperthermophile organisms in bold over their entire length. This drawing suggests that the most distant ancestors of current microbes from hell (at the beginning of the branches) were also hyperthermophiles.[20] In figure 4.6, we further observe that the branches linking LUCA to the common ancestor of archaea and to the common ancestor of bacteria are also drawn in bold, suggesting that LUCA was also a microbe from hell. Finally, the basal part of the tree plunging toward the origins of life is also drawn in bold, implying a hot origin of life.

New, Controversial Scenarios for a Hot Origin of Life

As we saw in chapter 1, the idea that life would have appeared in high temperatures, promoted by Woese and Stetter, went very well with the way in which scientists represented primitive Earth: an infernal world where volcanic activity was much greater than today and where the surface of our planet was constantly bombarded by raining meteorites. It is tempting to imagine that life appeared in heat at the bottom of the oceans—one of the favorite

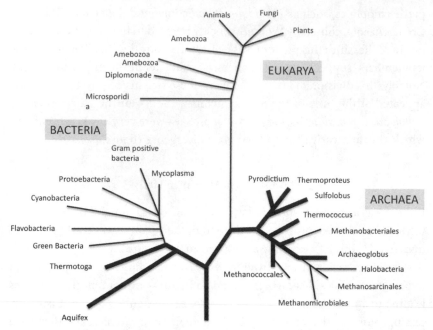

habitations of our microbes from hell—and even perhaps in Earth's crust, sheltered from cosmic rays (in particular, ultraviolet) and the destructive impact of the meteorite bombardment. The idea that LUCA was a microbe from hell was thus enthusiastically adopted by part of the scientific community, particularly those who were studying hyperthermophiles because, in those conditions, they found themselves at the heart of the fascinating problem of the origins of life.

Only a few isolated voices were heard contesting this scenario. Stanley Miller, who in 1953 had produced some amino acids in the lab by mimicking the supposed conditions of primitive Earth, and his good friend, Antonio Lazcano, professor at the University of Mexico, didn't support at all the idea of a hot origin of life.[21] For them, high heat was incompatible with the slow accumulation of organic molecules in the primordial ocean, an indispensable stage in Miller's model of the primordial soup. According to him, these molecules were too unstable at high temperatures; they would have disappeared on a burning primitive Earth well before they would have been able to be assembled into more complex structures.

At the same time, Miller's scenario of a primordial soup was itself being strongly contested. In 1987 Günter Wächtershäuser, a German lawyer who was also trained in chemistry and who worked, like Albert Einstein, at the Vienna patent office, proposed a "revolutionary" model to explain the appearance of life on Earth. According to Wächtershäuser, the first organic molecules were not formed in the atmosphere before accumulating in the primitive oceans, but would have been directly manufactured by chemical reactions on the surface of minerals of pyrite composed of iron sulfite.[22] He believed that everything had begun through the reduction of carbon gas (CO_2) by hydrogen sulfide (H_2S) and iron at a reductive state present in pyrite (FeS_2). The organic molecules synthesized by this reaction, at the beginning very simple, would have formed a film on the surface of the pyrite. They would have continued to evolve there, thanks to geothermal energy, giving birth to increasingly complex molecules, up to the macromolecules of the living being.

For Wächtershäuser, hyperthermophiles—in particular those that in chapter 2 we called stone eaters (chemolithotrophs)—appeared as direct descendants of these "primitive chemical beings" that lived on the surface of pyrite minerals. Given all this, it is not surprising to learn that he was a great admirer of Woese. Wächtershäuser thought that his new theory on the origins of life constituted the logical complement to the work that had led to enrooting the universal tree in a burning world. Not surprisingly, he was among the scientists invited by Woese himself to attend the Crafoord Prize ceremonies in Lund and Stockholm. The partisans of a hot origin of life were comforted by Wächtershäuser's theory, and most of them quickly adopted it. He even worked for some time in the 1990s with Stetter, who had put part of his lab at Wächtershäuser's disposal, to demonstrate experimentally the reality of some of the chemical reactions predicted by the new theory.[23]

Some authors proposed variations on Wächtershäuser's theory, which were distinguished from it by the type of chemical reactions implicated at the very beginning of life. A common point in all these hypotheses was the idea that the first living beings fed only on gases and on minerals to manufacture their organic molecules, as many contemporary hyperthermophiles do. All the variants also postulated that life had appeared in a mineral environment capable of protecting the first organic molecules from thermal degradation. One of the most popular was proposed by Michael Russell, a scientist from the Jet Propulsion Laboratory in Pasadena, California. In this scenario, life had appeared at the interface between an acidic hydrothermal fluid and the primitive ocean assumed to be basic.[24] This acid/base gradient would have engendered a flow of protons enabling the fixation of carbon oxide, which would have prefigured the flow of the proton that feeds modern

ATP synthases. At the end of the 2000s, Russell was able to obtain funding to reconstitute a reactor in the lab replicating a hydrothermal chimney; he hoped to be able one day to demonstrate the validity of his model.

An Alternative Hypothesis: A Cold Origin of Life and a Burning LUCA

The proponents of a "cold" origin, not surprisingly, were not convinced by the scenarios proposed by Wächtershäuser or Russell. They had first pointed out that Wächtershäuser, even associated with Stetter, had been able to reproduce in the lab only the very first reactions predicted by his theory, and even then with very poor yield. Similarly, the first results obtained with Russell's reactor are rather limited.[25] In any event, there still exists an enormous gap between the prebiotic chemistry directed by geochemistry that these authors are proposing and modern hyperthermophiles. The partisans of a "cold" origin would thus counterattack on the field of logic. Several big names in the origin of life and the RNA world fields, including Miller, pointed out that the hot LUCA of Stetter does not necessarily imply a hot origin of life. In a paper published in *Science* in 1995, they declared that life could very well have appeared in the cold and then evolved toward a hot or burning LUCA[26] (figure 4.7a).

To understand this argument, we must remember that LUCA should not be confused with the first cell that appeared on Earth. LUCA was already a rather evolved organism, since all current living beings (its descendants) inherited from it the same genetic code and the same complex mechanism of synthesis of proteins (with ribosomes and its constituents, including our famous 16S ribosomal RNA). LUCA was thus the product of a long evolution since the appearance of life (long in terms of the many evolutionary innovations that took place during this period, because we don't know at all how much time it took to get from the origin of life to LUCA). The weather thus had time to change several times from the very beginning of life to the time of LUCA.

Where We Must Take into Account the RNA World

I began to be interested in the origin (hot or cold) of life at the beginning of the 1990s, when I started studying the behavior of DNA molecules at high temperature, the results of which we discussed in the previous chapter. While analyzing the literature that reported on the experiments carried out in the 1960s regarding the stability of nucleic acids (DNA and RNA), I had been struck by the fragility of RNA at temperatures in which hyperthermophiles

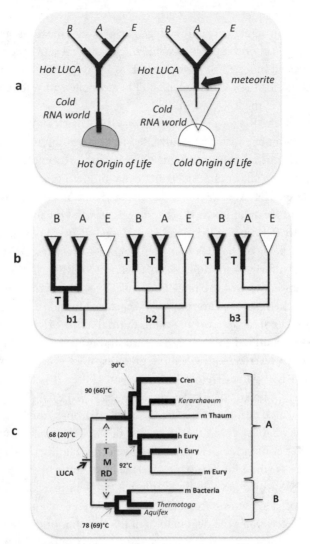

FIGURE 4.7. The origin of hyperthermophiles. Mesophiles are connected by thin lines, and hyperthermophiles by thick lines. *A*, on the left, a hot origin of life and a hot LUCA separated by a cold RNA world; on the right, a cold origin but a hot LUCA (from a lineage that survived a big meteoritic bombardment). *B*, three versions of the thermoreduction (T) hypothesis. In b1 and b2, the root is located between eukaryotes (E) and all other organisms; in b3, the root is located between bacteria and all other organisms. In b1, thermoreduction (T) occurred in a lineage common to archaea (A) and bacteria (B). In b2 and b3, thermoreduction (T) occurred twice independently after separation of the archaeal and bacteria lineages. *C*, schematic tree of Gouy and colleagues. Numbers are average predicted temperatures in Groussin and Gouy (2011). Numbers in brackets are average predicted temperatures in Boussau and colleagues (2008). The margins of error in both cases are large, varying from 10 to 30°C. Note also the differences between the predicted temperatures between 2008 and 2011. This indicates that these numbers should not be taken at face value, but as tendencies suggesting evolution from a mesophilic LUCA to a last common thermophilic ancestor for bacteria and a last common hyperthermophilic ancestor for archaea.

live. We saw in chapter 3 that RNA is particularly sensitive to exposure at high temperatures, owing to its reactive oxygen on 2' of the ribose and the absence of any topological bond capable of stabilizing its structure in space. If life had indeed appeared in a hellish environment, as I, myself, thought at the time, this characteristic of RNA seemed difficult to reconcile with a solid evolutionary theory, that of the RNA world.

What does "the RNA world" mean? It was the American Walter Gilbert, a Nobel laureate in chemistry, who invented this expression in 1986, although biologists are not all in agreement when it comes to explaining what it means.[27] However, all (or almost all) agree that modern organisms, with their genome made of DNA, were preceded during evolution by organisms whose genome was made up of RNA. This hypothesis was first put forward by Woese in 1967, rapidly followed by Crick and Leslie Orgel.[28] There are two very strong arguments in favor of the idea that RNA genomes preceded DNA genomes. First, DNA is a form of modified RNA (a variety of RNA, in a sense), since the sugar that composes the nucleotides of DNA (deoxyribose) is formed through the chemical transformation of the sugar found in the nucleotides of RNA (ribose). It is therefore logical to think that the original form (RNA) preceded the modified form (DNA). Also, today we have good reason to think that current enzymes, which are proteins, were preceded by enzymes made up of RNA, which were called ribozymes. Thus it is RNA, not DNA, that would have "invented" proteins (see box 4.2).

If life appeared at high temperatures and if microbes from hell are direct proof of this, we must admit that the phase of the RNA world also unfolded from the start at high temperatures. But given the fragility of RNA in those conditions, how could the ribozymes and the RNA genes of organisms populating the RNA world have endured infernal temperatures? Especially since the activity of ribozymes necessitates the presence of magnesium, whereas this element accelerates the degradation of RNA even more at high temperatures.[29] Furthermore, it is difficult to imagine that the first organisms with RNA had time to "invent" sophisticated strategies that modern hyperthermophiles develop to protect their RNA. We saw in chapter 3 that one of these tactics consists of blocking the oxygen on 2' of ribose by a methyl group to prevent breaking the link between nucleotides. This oxygen is precisely what ribozymes need to catalyze chemical reactions. So was the RNA of the first organisms protected by a mineral environment that excluded the water molecules necessary for thermal degradation? This is unlikely because without water, there is no life. In short, for many biochemists an RNA world at very high temperatures, with RNA cells floating in the pools of boiling water or in the cracks of a hydrothermal chimney bathing in a burning fluid, is quite

Today, all cellular living beings have a genome made up of DNA, but we are almost sure that this DNA world was preceded by a period when the genome of cellular organisms was made up of RNA. This theory, proposed in 1967 by Woese, henceforth rests on a very solid foundation, in particular with the discovery, at the beginning of the 1980s, of RNA molecules capable of catalyzing chemical reactions: ribozymes.[1] These true RNA enzymes contributed to the resolution of a problem that had long seemed unsolvable: which appeared first, DNA or proteins? You need DNA (genes) to manufacture proteins (it is DNA that determines the sequence of proteins in amino acids) and you need proteins to produce DNA (in particular, enzymes that synthesize new strands of DNA, called DNA polymerases). Biologists were thus confronted with a molecular version of the dilemma of the chicken and the egg: to make an egg you need a chicken (namely, a hen); to make a chicken you need an egg. With ribozymes, the situation was partially resolved: RNA can both behave as a protein, by acting as an enzyme, and play the role of DNA, by being the support of coded genetic information in the form of a succession of nucleotides (even today, the genomes of numerous viruses—AIDS, the flu, or even Ebola—are made up of RNA). Our DNA world would thus have been preceded by an RNA world where the only macromolecules of living beings were RNA: some playing the role of genes, and others, the role of enzymes. Later, RNA would have "invented" proteins and we would have entered into a new era, that of a world of more complex cells composed of RNA and proteins, but still not having DNA. Finally, DNA would have appeared thanks to the invention of protein enzymes capable of modifying the nucleotide substrates of RNA (ribonucleotides rA, rG, rC, rU) into deoxynucleotides, substrates of DNA (dA, dG, dC, dT). The model of the RNA world was established in 2000 by a fundamental discovery when crystallographers were able to decode the spatial structure of the great subunit of ribosome.[2] They were able to see in this structure the active site where the amino acids that form the proteins are assembled. This active site is made up uniquely of RNA (figure 4.2). It is thus the large molecule of ribosomal RNA (23S) that catalyzes the formation of the peptide link, explaining why I mention in chapter 3 that ribosomes are also ribozymes. As a result, since proteins as we know them (I used to call them modern proteins) are manufactured by an RNA molecule, they could not have appeared before RNA. This does not exclude that polypeptides or protoproteins existed before the emergence of the ribosome, but in that case, they should have been synthesized by other, now extinct, mechanisms.

1. For seminal reviews of the RNA world and ribozyme concepts, see Joyce, 2002; Cech and Bass, 1986. For the discovery of ribozymes, see Kruger et al., 1982; Guerrier-Takada et al., 1983.
2. Cech, 2000.

improbable. This was my opinion; the problem posed by the improbable existence of an RNA world at high temperatures led me to side with the partisans of a rather cold origin, a viewpoint that I discussed in several papers, including a review in the journal *Cell* in 1996.[30]

I had another argument against the idea of a direct link between a hot origin of life and modern hyperthermophiles, which I worked on every day in the lab. In my opinion, the complexity of their molecular biology suggested that these microbes corresponded to marvels of adaptation rather than primitive organisms. In the previous chapter we saw how microbes from hell protect their macromolecules—RNA, but also DNA and proteins—against the effects of temperature. If microbes from hell are direct descendants of the first forms of life on our planet, one must imagine that what we called tricks for adapting to high temperatures in fact correspond to ancestral characteristics. This doesn't seem plausible. The specific elements that we observe in macromolecules of archaea or bacteria living at very high temperatures cannot easily correspond to primitive characteristics. In all likelihood they are modifications of preexisting structures. For example, the chemical modifications of transfer RNA in microbes from hell are the result of complex enzymatic systems that certainly didn't exist when transfer RNA appeared. In the next chapter we will see that the study of the reverse gyrase was another decisive argument that convinced me that microbes from hell did not descend directly from the first forms of life that appeared on our planet.

The rejection of a direct link between a hot origin of life and modern hyperthermophiles did not, however, negate the possibility of a hot origin, followed by a cold period at the time of the RNA world (figure 4.7a, left panel) or the possibility of a direct link between modern microbes from hell and a hyperthermophilic LUCA that might have emerged from a cold origin of life followed by a cold RNA world (figure 4.7a, right panel). LUCA's ancestors indeed had time to develop adaptations enabling the RNA to endure high temperatures. But this raises a new question: If life was indeed in a cold stage at the moment of the RNA world, why would LUCA have been a microbe from hell, as suggested by the tree redrawn by Stetter? Gogarten proposed an audacious scenario in 1995 that brought back the explanation commonly advanced to explain the disappearance of dinosaurs—a cataclysmic scenario.

A Cold Cradle and a Hot LUCA? A Matter of Bombardments

According to Gogarten and his colleagues,[31] life appeared in the cold, then living beings were diversified and hyperthermophiles appeared. The latter occupied all the hot biotopes of the planet, in particular the deep oceanic hydrothermal springs. Everything was fine until the arrival of giant meteorites, whose size would have turned the meteorites that probably contributed to the disappearance of the dinosaurs at the end of the Cretaceous period (65 million years ago) green with envy. The shock would have been so violent that the entire surface of the planet (and perhaps even part of the oceans) would have immediately been sterilized. Only the cleverest of creatures would have survived: those who had taken the precaution of settling in the ocean depths, in the chimneys of our white or black smokers, which are so dear to microbes from hell. As all the rest of the living world would have perished, these heat-loving survivors became our ancestors by default (figure 4.7a, right panel).

This scenario rests on an event that actually occurred 3.9 billion years ago, which we call "the Late Heavy Bombardment" to distinguish it from the continual bombardment of meteorites that occurred when Earth was formed 4.5 billion years ago. If we look up into the sky when the moon is full, we can see, if our eyes are good, traces of this cataclysmic event that effectively shook the solar system 3.9 billion years ago.[32] The lunar seas and the giant craters we can observe on the surface of our satellite are proof of this gigantic bombardment of meteorites that did not spare our planet. Could life have survived such a pounding? This is a mystery. If yes, then, according to Gogarten, the bombardment would have killed off all the nonhyperthermophile contemporaries of LUCA, leaving only our presumed ancestor.

In my opinion, this scenario still has a weak point. It assumes that the undersea smokers—the only inhabited spaces protected from the bombardment—were inhabited only by hyperthermophiles. But as we know, this is not the case. The walls of the smokers also shelter many moderate thermophiles or even mesophilic bacteria that reside closer to the exterior face of the chimneys, which is near the cold water (between 2 and 4°C) of the ocean floor.[33] The smokers of the oceans where LUCA and its contemporaries lived must have also been populated with varied fauna, including organisms adapted to temperatures that were extremely hot or extremely cold. In those conditions, why would hyperthermophiles have been the only ones to survive the sterilization of the planet? We must imagine that the Late Heavy Bombardment liberated enough energy to heat the water of all the oceans to a temperature close to the boiling point of water, but not hot enough to eradicate every trace of life on Earth. This scenario seemed highly improbable to me, and the question "Why was LUCA a hyperthermophile?" still hadn't been answered. Ultimately, I came to wonder whether LUCA had truly lived in hell. If it hadn't, could Stetter's tree, with its bacterial root surrounded by hot organisms, archaea and bacteria, be explained?

Doubts about the Root

To resolve the dilemma of a cold LUCA, suggested by considerations of a theoretical nature, and of a hot LUCA, suggested, it seemed, by the bacterial root of the universal tree, it was simply a matter of imagining another root. More precisely, one had only to situate the root between eukaryotes (E) and a common ancestor of archaea (A) and bacteria (B; figure 4.7b, b1). In that case, one could imagine a mesophilic LUCA coming from a cold origin of life, and a hot common ancestor of archaea and bacteria. I really liked that root, because it enabled me to imagine a LUCA that would have had a certain number of "primitive" characteristics that would have persisted in modern eukaryotes, but would have disappeared in archaea and bacteria—for example, genes in pieces, which we discussed in chapter 2. I had the impression that this configuration of the tree would have definitively refuted the old idea of evolution going from the simple to the complex, from prokaryotes to eukaryotes. Of course, Woese detested this root, because it again brought archaea and bacteria together. This was one of the points on which we differed, which did not prevent us from having a cordial discussion during the Crafoord Prize festivities.

Fine, you might say, but if we've understood correctly, the question had been settled by the work of Iwabe and Gogarten, who had concluded unambiguously in favor of the bacterial root (figure 4.5b). Exactly, but I wasn't convinced at that time that their work was beyond argument. The principle of the method they had used was relatively simple, but its application should be viewed with caution: as we have seen, it implied the comparison of very ancient proteins by going back in time, even before the time of LUCA. Phylogenies obtained in these conditions could be biased by a certain number of artifacts, in particular if the branches leading to modern organisms don't have the same length. If we look again at figure 4.5b, we see that in both subtrees (those of EFTu and EFG), the branch leading from the point connecting the three domains (noted by an arrow) to bacteria is much longer than those connecting the same point (I call it the trifurcation point) to the two other domains. This is also the case in the trees of ATP synthases.[34] This observation has led some scientists, including me, to assume that the rooting of these trees between bacteria on the one side and archaea/eukaryotes on the other was possibly due to an artifact well known to phylogeneticists: the long branch attraction artifact. To understand what it means, it's probably better to call it "short branch attraction." Let's recall that the length of the branches on our trees is more or less proportional to the number of mutations that have accumulated during evolution. The sequences of organisms located at the end of the short branches have thus undergone fewer mutations than those of organisms located at the end of the long branches. Consequently, two organisms at the ends of short branches will have conserved more ancestral amino acids in their sequences than those corresponding to the long branches. These sequences will resemble each other because of the preserved ancestral amino acids; the computer programs of the construction of the tree will group them together even if this doesn't correspond to historical reality.

We have seen that the phylogenetic trees obtained from the sequences of nucleic acids or amino acids are broadly constructed on the principle of "that which resembles assembles." This principle can be deceptive when one seeks to reconstitute the history of organisms: for example, the chimpanzee resembles a gorilla more than it does a human, at first glance. However, evolutionists have shown that the human is the brother of the chimpanzee and the cousin of the gorilla. If the chimpanzee resembles the gorilla more than the human at first glance, it is because they share a greater number of ancestral characteristics (present in the ancestor of the three groups) that have been modified or have disappeared more recently in humans. By contrast, the human and the chimpanzee share a certain number of new characteristics (derivative characteristics) that prove a lineage common to these two groups of great apes (and yes, we are great apes). To achieve good reconstructions of the history of a group of organisms, one must be able to distinguish between ancestral and derived characteristics to use only the latter to classify organisms. Indeed, ancestral characteristics may have been lost independently in different lineages, confusing the picture. This very simple strategy, called cladistics analysis, was proposed by a German specialist of insect evolution, Willi Hennig. It is very difficult to implement in the evolution of sequences because very few positions in the alignment of nucleotides or amino acids lend themselves to this type of analysis.[35]

I had discovered the principles of cladistics analysis while reading Stephen Jay Gould's book *Wonderful Life*, which describes how paleontologists reconstituted the history of fossils of ancient animals discovered in the Burgess shale in Canada. By applying cladistics analysis to the sequence alignments of the elongation factors EF1 and EF2 published by Iwabe, I realized that these alignments didn't contain any significant signal enabling me to determine where the root of the universal tree should be located. I had tried in vain to publish these results in *Nature*. Finally, they appeared in 1993 in *Biosystem*.[36] I then devoted myself to this problem in collaboration with Hervé Philippe, a young scientist who worked at Orsay at the time and whom I considered to be the world specialist in molecular phylogenetics. The computer analyses he undertook were to confirm the analysis I had done by sight and by hand on the sequences of elongation factors, then on other protein pairs used by several scientists to root the universal tree. In every case, the results showed that it was impossible to know whether the root obtained for the branch leading to bacteria was owing to an artifact or corresponded to reality.[37] Skeptical, I couldn't see—nor did Hervé—why the root couldn't be localized on the branch of eukaryotes, which enabled me to imagine a cold LUCA and a hot ancestor common to archaea and bacteria, explaining the grouping of hyperthermophiles around the point of divergence between these two prokaryote domains (figure 4.5b). This scenario of course raised a new question: Why, if LUCA lived in the cold, was the ancestor common to the two prokaryote groups a microbe from hell? How can we explain the puzzling emergence of a common hot ancestor of archaea and bacteria from a cold LUCA?

A New Scenario for the Origin of Microbes from Hell: Thermoreduction

While thinking about this problem, I found a possible solution that was particularly attractive to me: an adaptation to increasingly high temperatures would have led to the appearance of the prokaryote phenotype of the ancestor common to bacteria and archaea. In this scenario, the small size and the rapid turnover of macromolecules, characteristic of prokaryotes, would have been selected as favorable traits during the adaptation of their ancestors to higher and higher temperatures. Indeed, rapid macromolecular turnover, facilitated by a small size, enables rapid replacement of macromolecules damaged by thermal degradation. This was especially important for RNA. In this scenario, LUCA could have thus been a more complex organism than modern prokaryotes. Perhaps it already possessed a nucleus? The loss of this nucleus would have enabled the messenger RNA to transmit its message before being destroyed by temperature through the coupling between transcription and translation, characteristic of prokaryotes.

After having tried several prestigious journals such as *Nature*, I managed to publish my so-called thermoreduction hypothesis in 1995, in the *Comptes Rendues de l'Académie des Sciences*, a French journal that unfortunately has a small circulation, even though it is published in English.[38] In the same paper I presented my idea, discussed in chapter 3, according to which the fragility of RNA explained why eukaryotes were incapable of living at temperatures above 60°C. Also in this article I predicted that we would probably not find hyperthermophile eukaryotes, due to the instability of RNA at high temperatures, given the necessity for eukaryote messenger RNA to be sufficiently stable to endure their transport from the nucleus to the cytoplasm. We have seen that this prediction holds for the moment. This is not for lack of trying to find hyperthermophile eukaryotes. If you remember from chapter 1, Stetter himself was unable to find eukaryotes living at higher than 55°C. Moreover, by using methods of molecular ecology, Norman Pace did not succeed in detecting eukaryotes in the hot springs of Yellowstone by using as probes fragments of eukaryotic 18S ribosomal RNA. The thermoreduction paper had little impact except among a few evolutionist colleagues, such as Nicolas Glansdorff in Belgium and David Penny in New Zealand, who shared my point of view that prokaryotes should not be considered primitive organisms, and who referred to my theory in a few papers.[39]

In my article I had taken care to emphasize that, ultimately, my hypothesis could be equally valid if the root of the universal tree were indeed on the branch of bacteria (figure 4.7b, b2). It was necessary to imagine, in that case, two independent thermal reduction events (T) leading on the one hand to bacteria and on the other to eukaryotes. At the beginning, this hypothesis appeared a bit complicated. But the more I reflected, the more I wondered if, after all, the hypothesis of a hyperthermophile ancestor common to archaea and to bacteria, suggested by Stetter's tree and adopted by most scientists studying microbes from hell, was as solid as it seemed.

Challenging the Hyperthermophilic Prokaryote Ancestor?

In the universal tree redrawn by Stetter, the hyperthermophiles are grouped around the root, and their branches are generally shorter than those leading to mesophilic organisms (figure 4.6). This observation was interpreted as proving the primitive character of hyperthermophiles. They had evolved less since the era of the ancestor common to archaea and bacteria. But I ended up realizing that the grouping of microbes from hell around the root, as well as their short branches, could simply be explained by the wealth of their 16S ribosomal RNA in nucleotides G and C. In the previous chapter we saw that this wealth enables the stabilization of the minihelices present in 16S ribosomal RNA (figure 3.6c). This wealth of ribosomal RNA in G and C has the effect of slowing the speed of evolution of the sequences of this molecule: in fact, many positions that evolve freely in the 16S ribosomal RNA of mesophilic organisms—a nucleotide C may be replaced by an A, a U, or a G—will necessarily be occupied by a G or a C in hyperthermophiles. In other words, the space of sequences that can be explored by their 16S ribosomal RNA during evolution will be more restricted than that accessible to 16S ribosomal RNA of mesophilic organisms. Clearly, the 16S ribosomal RNA of microbes from hell cannot accumulate as many mutations as those of mesophiles. Consequently, since their 16S ribosomal RNA evolves more slowly, microbes from hell will automatically be located at the extremity of shorter branches than the other organisms in the tree based on 16S ribosomal RNA. Also, in two hyperthermophiles, the same position may be occupied by a G or a C, not because this nucleotide corresponds to a primitive state that was inherited from a common ancestor, but through convergence: the two organisms may have independently selected a G or a C in this position because it better stabilizes the molecule at high temperatures. The result: the 16S ribosomal RNA of two hyperthermophiles can end up resembling each other, even if they are not specifically related. The construction of phylogenetic trees being based on the principle that "what resembles assembles," hyperthermophiles—archaea and bacteria—will be attracted to the base of the universal tree of life by their abnormal composition of bases G and C. Thus it is difficult to know if the grouping of hyperthermophiles around ancestors of bacteria and eukaryotes corresponded to historical reality or simply to this methodological bias. And so how can we now choose between a hot or cold environment for the ancestor common to the two prokaryote domains?

The comparison of the lipids of hyperthermophile archaea and bacteria provided an argument that seemed to counter the hypothesis of a burning ancestor of archaea and bacteria. As we saw in chapter 3, the lipids of archaea (archaeol) are very particular: they have thick hydrophobic carbon chains connected to the carbon skeleton of glycerol by two ether (not ester) bonds, which gives them great solidity and enables their membranes to remain stable and impermeable at high temperatures. The lipids of hyperthermophilic bacteria are also adapted to high temperatures, but are very different from those of archaea. Granted, they resemble archaeal lipids in some ways, since a few have solid heads (ether bonds) and others can apparently form giant lipids from "diabolical" acids. But these are typically bacterial lipids: they do not have the thick carbon chains (branched isoprenoids) typical of archaeal lipids, but fatty linear acids (or alcohols); more important, they have the same stereochemistry as all other bacterial lipids (figure 3.9). If the ancestor common to bacteria and archaea had been a microbe from hell, then in all likelihood it would have bequeathed its lipids adapted to life at high temperature to the ancestors of *all* contemporary hyperthermophilic organisms, either bacteria or archaea. In these conditions, we should still be able to observe the *same* type of lipids in *all* modern hyperthermophiles. Instead, we see that bacteria from hell have typically bacterial lipids, even if on some points they resemble archaeal lipids. These resemblances correspond to a classic phenomenon of evolutive convergence.

In summary, everything indicates that the adaptation of lipids to high temperatures unfolded independently in archaea on the one hand and in bacteria on the other, starting with lipids "foreseen" at the beginning for lower temperatures. By adapting to high temperatures, these lipids of archaea or bacteria preserved their fundamental differences while acquiring a few common characteristics through convergence. If this reasoning is correct, it means that the ancestor common to bacteria and archaea was not a hyperthermophile. In that case, regardless of the position of the root of the universal tree, the hypothesis of thermoreduction indeed involved two independent events of adaptation to hot conditions from a cold LUCA. I had thus lost a bit of faith in this hypothesis, which was no longer as simple as it had seemed, until the day when some colleagues in Lyon obtained results that restored my faith. We will meet them now in the large auditorium of the museum of primitive arts in Paris for another awards ceremony.

October 23, 2012: Musée du Quai Branly, Paris

The Musée Branly exhibits magnificent collections of African, Oceanian, Asian, and Amerindian art in an avant-garde building set on the banks of the Seine, very close to the Eiffel Tower. Its auditorium is often used for scientific or political events (the oath of the minister of research, for example), and each year it also serves as the venue for the ceremony bestowing awards for the best scientific papers from France published in the preceding year, sponsored by the journal *La Recherche*, the French equivalent of *Scientific American* or *The Scientist*.

This year I was invited to this ceremony because the laureate for biology is Manolo Gouy. This scientist from CNRS who works in Lyon climbs onto the stage to receive his prize from the master of ceremonies. Gouy is accompanied by his PhD thesis student, Mathieu Groussin, who was the first author of the best paper of the year in fundamental biology. For me, the award recognizes work that Gouy and his colleagues undertook at the end of the 1990s and that had a great impact on our view of LUCA. They had asked this question: How can we determine the temperature at which LUCA lived? Their idea was to focus on a characteristic of microbes from hell that we have just been discussing: the wealth of their ribosomal RNA in base pairs GC. The scientists' reasoning was as follows: if we manage to determine the percentage of G and C nucleotides in the 16S and 23S ribosomal RNA of LUCA, we could, through comparison with current organisms, predict whether LUCA lived at low or high temperatures. If this proves that the ribosomal RNA was rich in G and C, with a percentage of GC higher than 60%, which is the case for the ribosomal RNA of hyperthermophiles, this would constitute an argument in favor of a LUCA having lived in hell. In the opposite case, if the percentage is between

50% and 60%, typical of the ribosomal RNA of mesophiles, we can confirm that LUCA didn't live in infernal temperatures. How to determine the base G and C percentage in the ribosomal RNA of LUCA? We can't ask it to provide us with its RNA, since it hasn't existed for a long time. However, it is possible to tentatively reconstruct the ancestral sequence of a macromolecule from the current sequences of this same molecule. Gouy, originally a mathematician, had developed analytical methods that enabled him to determine the evolution of the percentage of the four bases of RNA or of the twenty amino acids of any protein, along the nodes and branches of a phylogenetic tree. It was then possible to apply this method to the analysis of the universal tree of life, focusing on its more important node, LUCA.

In 1999, Gouy and his colleagues published the results of their first analysis in the journal *Science*. The verdict: the percentage of G and C nucleotides in the ribosomal RNA of LUCA must have been around 55%, which corresponds to a mesophilic organism.[40] The significance of these results was, however, limited at the time by the still relatively low number of ribosome sequences available for analysis. And so they resumed the study a few years later, once they had a much greater number of sequences, covering new groups, which they hadn't been able to study in the earlier work. In addition, they were able to extend their study to the proteins of LUCA. In the meantime, the genomic revolution (which we will discuss in more detail in the next chapter) had taken place. Many genomes of organisms belonging to the three domains of life had been sequenced, and we had been able to deduce their "proteome"— that is, the entirety of the proteins coded by the genes present in these genomes. The scientists were able to identify a few dozen universal proteins that are present in all living beings and therefore were certainly present in LUCA. In parallel, scientists were also now able to compare the proteomes of mesophilic and thermophilic organisms for the composition of their proteins in amino acids. They had observed that the hydrophobic amino acids valine (V), isoleucine (I), leucine (L), and tryptophan (W)—as well as two amino acids carrying an electric charge, aspartic acid (E, negatively charged) and arginine (R, positively charged)—are more abundant in hyperthermophilic than in mesophilic proteins. By contrast, other amino acids are more abundant in mesophilic proteins, such as histidine, glutamine, or asparagine.

If you read the previous chapter carefully, you should understand why amino acids VILWER favor the formation of ionic (ER) and hydrophobic (VILW) bonds that strongly stabilize the structure of proteins at high temperatures. They will thus accumulate in the sequences of thermophilic proteins to the detriment of other amino acids—those that, on the contrary, favor the flexibility of proteins and thus destabilize its structure at high temperatures. We can then deduce from this a correlation between the temperature at which an organism lives and the composition of its proteins in amino acids. Once again, by reconstituting the ancestral sequences of universal proteins in LUCA, it became theoretically possible to determine at what temperature our ancestor lived from the amino acid composition of its proteins.

A Cold LUCA and Two Hot Ancestors

The analysis of universal proteins by Bastien Boussau, a student of Gouy's, confirmed the earlier results obtained with ribosomal RNA. The composition in amino acids of the proteins of LUCA was characteristic of that of meso-philic proteins. From this analysis, LUCA lived at temperatures lower than 40°C. These new results, published in *Nature* in 2008, reinforced the idea that LUCA had not lived in hell.[41] But the results presented in this paper went even further. The method used enabled the authors to determine the temperature at which not only LUCA but also all the other intermediary ancestors that had lived between LUCA and us had lived: for example, the last ancestor common to archaea and the last ancestor common to bacteria. And this was a big surprise. Whereas LUCA (from the analyses) had lived at low temperatures, the last common ancestor of all archaea and the last common ancestor of all bacteria had both lived at high temperatures, around 70°C (figure 4.7c).

In the 2011 article, which was published in *Molecular Biology and Evolution* and won an award from the journal *La Recherche*, Groussin had repeated this study with an even greater number of protein sequences and confirmed the surprising results.[42] In this new analysis, LUCA was now a moderate ther-mophile, since its presumed optimal growth temperature was 65°C. By con-trast, the last common ancestor of archaea was now even more hyperthermo-phile than in the 2008 article, with an optimal growth temperature around 90°C. This was now clearly higher than that of the last common ancestor of bacteria, which was around 75°C. Of course these numbers should be taken with a grain of salt, because the margins of error aren't small.

Notably, a particularly hot ancestor for archaea (90°C) fits perfectly with the fact that archaeal lipids are so well adapted to extremely high tempera-tures. One could imagine that these lipids were selected during the evolution period that took place from LUCA to the last common ancestor of archaea (archaeogenesis), allowing them to explore new, increasingly hotter biotopes, up to a maximum of 110°C. However, this point is still debated, because the nature of lipids in LUCA remains a mystery. I think that LUCA had bacterial/ eukaryotic-type lipids in its membrane (in line with its mesophilic nature), but other authors think they were of archaeal type (isoprenoids), while still others think that LUCA had two types of lipids and that the "great lipid divide" took place when the separation of the archaeal and bacterial lineages occurred.[43]

In any case, as you have most certainly guessed, I was delighted at the results Gouy and his students obtained because their conclusions, two adaptations independent of the hyperthermophile nature, went perfectly with my thermoreduction hypothesis for the origin of prokaryotes, archaea, and bacteria (T in figure 4.7c). For me, this was close to being an experimental confirmation of the thermoreduction hypothesis. As we have seen, I had anticipated in my 1995 paper two independent stages of thermoreduction if the bacterial root of the universal tree was correct (figure 4.7b, b3). This fits perfectly well with the results obtained by Gouy and his students. And so I think that the thermoreduction hypothesis should be taken more seriously by the scientific community. Let's wait and see and not declare victory too soon. The optimal temperatures for growth deducted from the study of universal proteins are still capable of moving in the future, depending on the new sequences available and new methods for analyzing them.

The idea of a hot origin of life and a direct link between that origin and microbes from hell had seduced the community of scientists studying hyperthermophiles because it gave hyperthermophiles considerable importance in the history of life on Earth. This hypothesis, moreover, is still in favor with many of them. I think, however, that the scenario of thermoreduction takes nothing away from the importance of microbes from hell. In fact, in this scenario, thermophiles and hyperthermophiles are at the origin of two of the three domains of life, those of archaea and bacteria, respectively, which is nothing to sneeze at. These bacteria and, to a lesser extent, the archaea today dominate all terrestrial ecosystems, including our own ecosystem, in the form of the human microbiome. Also, in the thermoreduction scenario, thermophiles are indirectly at the origin of modern eukaryotes, since the latter all possess mitochondria of bacterial origin in their cells. Finally, thermophiles are indirectly at the origin of the oxygen present in our atmosphere, since it is produced by cyanobacteria, free or present in the leaves of plants in the form of chloroplasts. Without microbes from hell, life on Earth would have thus been very different from what we know today.

We can, however, imagine other hypotheses to explain the results obtained by Gouy and his colleagues. In their *Nature* paper they proposed two. In the first, they picked up the meteorite scenario that Gogarten had proposed to explain the existence of a hot LUCA after a cold origin of life. They simply moved the time of the Late Heavy Bombardment after the time of a cold LUCA (M for meteorite in figure 4.7c). This bombardment would thus have killed all the descendants of LUCA living at cold or warm temperatures, leaving alive only two lineages of thermophilic organisms: the ancestors of bacteria and those of archaea. I don't really like this explanation, for the same reason as before: that one must imagine that the late bombardment was sufficiently violent to make the bottom of the oceans around the black or white smokers boil, while sparing microbes from hell. In addition, in this scenario, eukaryotes must also be the descendants of the hyperthermophile survivors, which seems highly improbable to me. Am I right or wrong on this point? We will see in the next chapter that many evolutionists disagree with me on the scenarios for the origin of eukaryotes. Many of them tacitly support a hyperthermophilic ancestor of eukaryotes, because they think that we (eukaryotes) descend from archaea, which themselves descend from hyperthermophilic organisms.

In the second hypothesis, Gouy and his colleagues proposed that the passage from a cold LUCA to hot ancestors to archaea and bacteria was to be related to the transition of organisms having an RNA genome toward those possessing DNA genomes (RD in figure 4.7c). We saw in chapter 3 that DNA is much more stable than RNA at high temperatures because it lacks the reactive oxygen on 2′ of the ribose. DNA could thus have appeared in organisms during adaptation to conditions of life at high temperatures. These would have gradually eliminated all the less effective organisms whose genomes had stayed with RNA. In that case, LUCA would have itself been an organism of the RNA world. Note that this second hypothesis is not incompatible with that of thermoreduction because the passage from RNA to DNA could have been one of the major adaptations leading to the prokaryote phenotype. All the same, is the hypothesis of an RNA-LUCA plausible?

The Great DNA Replication Divide: Vista for an RNA-LUCA?

Remarkably, Woese and Fox had already proposed an RNA-based LUCA at the end of the 1970s in a paper published in the *Journal of Molecular Evolution*.[44] In this paper they had advanced for the first time the idea that LUCA had been a progenote, a very primitive organism. According to them, this progenote would have still possessed an RNA genome. This idea did not, however, receive a lot of attention. Everyone thought that DNA had already replaced RNA at the time of LUCA, since all modern organisms of the three domains have a DNA genome and the basic mechanisms of replication of the double helix of DNA are the same in the three domains of life.[45] In particular, this involved three categories of enzymes present at the same time in bacteria, archaea, and eukaryotes: a DNA helicase, which separates the two strands of

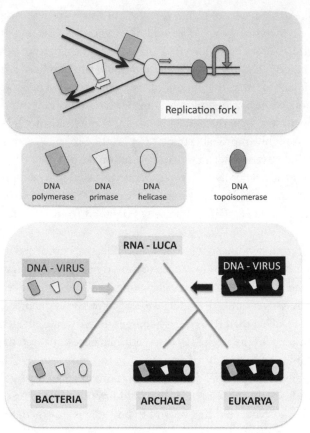

FIGURE 4.8. A schematic replication fork. A DNA helicase opens the two DNA strands. For structural reasons, DNA synthesis (*bold arrows*) proceeds in opposite directions on both strands, whereas the fork progresses in one direction (*gray arrow*). DNA polymerases require a small oligonucleotide primer to initiate DNA synthesis. A primase synthesizes first an RNA primer at the replication origin of the chromosome, which is used by the DNA polymerase that synthesizes the strand in the same direction as the fork. The primase then synthesizes RNA primers continuously on the strand, replicated backward for the polymerase working in the opposite direction. In one version of the "out of viruses" hypothesis (depicted here), two sets of DNA replication proteins (*gray and black boxes*) were transferred from viruses to cells after LUCA.

the molecule; a DNA primase, which primes the synthesis of the new DNA strands; and a DNA polymerase, which carries out this synthesis (figure 4.8).

Yes, but if LUCA already had a DNA genome, the enzymes replicating DNA today in modern organisms should be homologous in the three domains of life. In other words, each of them should be issued from an ancestral enzyme that replicated the genome of our LUCA. But this is not the case—and this is a major and puzzling result obtained from the genomic revolution. The first comparative analyses of genomes of organisms belonging to the three domains, carried out at the end of the 1990s by the team of Eugene Koonin, an American scientist of Russian origin whom we will get to know better in the next chapter, showed that the three major enzymes involved in DNA replication today are not present in the list of universal proteins present in LUCA. The DNA polymerase, the DNA primases, and the DNA helicases involved in archaeal DNA replication are related to those of eukaryotes, but not to those of bacteria. This is what I will henceforth call the great DNA replication divide, in reference to the lipid divide previously discussed. To explain this puzzling observation, Koonin and his colleague Arcady Mushegian proposed in 1996 that the mechanisms of DNA replication had been invented twice independently: once in the branch of the universal tree leading to bacteria, and once in the branch common to archaea and eukaryotes[46] (figure 4.8). This hypothesis thus brought back into fashion the idea of LUCA with an RNA genome, originally proposed by Woese and Fox. As we have just seen, this hypothesis fits perfectly with the hypothesis of Gouy and his students, suggesting two independent transitions of RNA toward DNA to explain the two independent transitions of a cold LUCA toward the two hot ancestors of bacteria and archaea. Note that the idea of an RNA-LUCA contrasts, once again, with that of a direct link between a potential hot origin of life and our microbes from hell. Indeed, RNA being fragile at high temperatures, it is difficult to imagine a hyperthermophilic LUCA with an RNA genome.

Although Koonin had been the first to revive Woese and Fox's idea of a LUCA with an RNA genome to explain the great DNA replication divide, he later opposed this idea, considering that RNA was not stable enough to enable LUCA to have a sufficiently large genome to be able to code all the proteins necessary for its metabolism.[47] As an argument, Koonin noted that the genome of the biggest RNA virus, the coronavirus (such as the SARS virus), did not go beyond the 32,000 base pairs, which enabled the coding of some forty proteins, far from the several hundred proteins expected for LUCA.[48] Another frequent assumption is that a large RNA genome cannot exist because, unlike DNA, RNA cannot be replicated very faithfully. I don't see why one should discard the RNA-LUCA hypothesis so easily. I think that some evolutionists have a tendency to underestimate RNA. Let's remember that RNA is sufficiently stable to enable the existence, among eukaryotes, of large RNA (primary transcripts) that can be the same size as a bacterial genome. The contemporary RNA viruses are not really good models of ancient RNA cells because the small size of their genomes, which enables a rapid replication, is a selective advantage for a virus. However, even RNA viruses can sometimes "speak" in favor of RNA. For instance, it has been recently shown that coronaviruses, such as the SARS virus, replicate their genome very faithfully using a rather complex replication machinery.[49] One should not forget that RNA viruses code for proteins as complex as those coded by DNA viruses. Finally, let's note that RNA cells that existed before DNA cells must indeed have been complex enough to manufacture the very sophisticated protein that catalyzes the transformation of RNA substrates, ribonucleotides, into DNA substrates, deoxyribonucleotides.[50]

You have probably noted that the idea of an RNA-LUCA to explain the great DNA replication divide implies placing the root of the universal tree between bacteria and archaea/eukaryotes (figure 4.8). You may remember that I was opposed to this root when it seemed to imply a hot LUCA. After the work of Gouy and his students, which showed that there could have been a cold LUCA and a bacterial root with two hot ancestors for archaea and bacteria, I no longer had a reason to disagree on a theoretical level.[51] I now think that the bacterial root of the universal tree of life is probably correct because, as we will see in the next chapter, it better explains some major differences between bacteria and the two other domains.

Let's return for the moment to the great DNA replication divide. This puzzling observation raises new questions: first, "How and why were the proteins involved in replication invented twice independently?" and more generally, "Where does DNA come from?"

Viruses Take the Stage

I have some ideas, again very personal, to answer the above two questions. In my opinion, DNA and the mechanisms of its replication first appeared in viruses, and it is viruses that then transferred DNA and DNA replication proteins independently in the lineage leading to bacteria and in the one leading to archaea and eukaryotes (figure 4.8). I published this hypothesis in 2002 in the journal *Current Opinion in Microbiology*.[52] I later modified it slightly by assuming three independent transfers of DNA from viruses to cells, one per domain, to explain a certain number of differences that exist among the mechanisms of replication in archaea and eukaryotes, in particular at the level of their DNA topoisomerases. To my greatest satisfaction, Woese, who had never been interested in viruses up to then, agreed to sponsor my paper for the *Proceedings of the National Academy of Sciences*, where it was published in 2006.[53]

How did I come to introduce viruses into a large-scale evolution scenario and into the origin of the macromolecule that one often identifies with life itself, DNA? Everything began with an unexpected discovery made in 1979 by Leroy Liu of the team of Bruce Alberts in the United States. These scientists had discovered that a well-known virus, the bacteriophage T4, possessed a type II DNA topoisomerase very different both from that of bacteria (DNA gyrase, see chapter 2) and from that of eukaryotes,[54] almost as if T4 represented a fourth domain of life (as you can see, I always return to my first loves, DNA topoisomerases). Since then, many scientists have discovered in the genomes of DNA viruses a large number of genes coding for proteins involved in the replication of their DNA: DNA topoisomerases, but also DNA primases, DNA helicases, or DNA polymerases. Remarkably, these viral proteins usually do not resemble (or resemble very distantly) the enzymes that accomplish the same function in the host cell of these viruses. Furthermore, these viral proteins are often also very different from one virus to another: those of bacteriophage T4 do not resemble those of the herpes virus, which do not resemble that of smallpox virus, and so forth. Thus in the current biosphere, besides the two different sets of cellular DNA replication proteins—those found in bacteria and those found in archaea and eukaryotes—scientists have discovered over the years several other sets of DNA replication proteins that are each characteristic of a different family of DNA viruses.

The "Out of Viruses" Hypothesis

Just as the diversity of genetic markers of African populations suggests an African origin of modern humans (this is the "out of Africa" hypothesis), the diversity of proteins of DNA replication in viruses, in my opinion, suggests a viral origin of these proteins (this is the "out of viruses" hypothesis for DNA and its replication). Europeans, Asians, and Amerindians possess only a subgroup of genetic markers present in African populations: those that the first *Homo sapiens* who came out of Africa possessed. Similarly, I think that the DNA replication proteins of bacteria and those of archaea/eukaryotes are two subsets of all DNA replication proteins that first appeared in viruses. They correspond to two groups of ancient viral proteins that enable the manufacture and replication of DNA and that were transferred, a long time ago, from viruses to cells. These transfers would have been responsible for the transition from cellular RNA genomes to cellular DNA genomes in the three domains of life.

If LUCA was indeed an RNA organism, as first suggested by Woese, that means that these transfers took place after the divergence of the lineages leading to bacteria and to archaea/eukaryotes. By contrast, if LUCA already possessed a DNA genome, we must imagine a first transfer of DNA replication proteins before LUCA, followed by the replacement of these proteins by nonhomologue proteins (functional analogs) either in bacteria or in archaea/

eukaryotes. I favored this scenario once,[55] but it now seems less probable, although it cannot be completely ruled out.

In the "out of viruses" hypothesis, we see that all the proteins that replicate the DNA of current cells would be of viral origin. How can this be explained? In my opinion, the simplest is to imagine that DNA itself appeared for the first time in viruses. This idea, understandably unexpected, is reasonable if one considers two things. First, the stability of DNA compared to RNA presented a specific selective advantage for viruses by enabling the production of virions capable of surviving longer in the environment, especially at high temperatures. Second, in the RNA world, a virus able to replace its RNA genome by a DNA genome becomes instantly resistant to attacks by cellular enzymes designed to destroy viral RNA. Such a mutant, the first organism with a DNA genome, must have had an immediate advantage in the competition between cells and viruses. Notably, some modern viruses still use this trick, chemical modifications of their nucleic acids, to bypass cellular defenses. This is the case, for instance, with the virus T4, which has a modified DNA genome in which a hydroxymethyl group has been systematically added to all cytosines.[56] This modification enables T4 to resist enzymes produced by bacteria to destroy its genome (these are the famous enzymes of restriction used by molecular biologists to cut DNA at precise places).

The descendants of these first DNA viruses must have diversified for a long time while giving birth to many lineages that would have all developed a great variety of increasingly complex mechanisms of DNA replication. The first DNA genomes could have been single stranded and replicated by a single protein, a DNA polymerase. Double-stranded DNA genomes would have appeared later on, requiring additional proteins, such as DNA primases and DNA helicases, for their replication (figure 4.8).[57] The DNA polymerase would have then become more complex to replicate more quickly and more faithfully by associating with new proteins.[58] DNA topoisomerase would have finally intervened to resolve the topological problems raised by the replication of circular DNA or of large linear DNA. A great variety of enzymes belonging to various protein families, RNA polymerase and RNA helicase, which until then worked on RNA, could have been recruited by different viruses to carry out all these functions. In the end, several replication mechanisms would have thus emerged in different viral lineages, explaining why today we discover a great variety of nonhomologous replication proteins in viruses.[59] Two of these mechanisms would have then been transferred from viruses to cells in the course of evolution—what we find today in bacteria, and what we find in archaea and eukaryotes—thus explaining the great DNA replication divide[60] (figure 4.8). This would have given birth to the first DNA cells, which would have rapidly eliminated all RNA cells. DNA, being more stable than RNA, enables the construction of bigger genomes, which can then carry more genetic information, which became a long-term important selective advantage for DNA cells.

It is usually advanced that DNA replaced RNA because it was a necessary step in the evolution of life toward organisms with large genomes, such as humans. But only an intelligent designer could have made such a prediction. Evolution does not work that way, as Darwin discovered. You have to explain the selective advantage of the first organisms that experienced such drastic change in their makeup, not the selective advantage of their descendants. We know today that dinosaurs had feathers. Scientists are still discussing different hypotheses about the role of these feathers (ornamental; to heat the body), but they all agree that feathers were not there just because dinosaurs at some time became birds! As we have seen, the "out of viruses" hypothesis for the origin of DNA provides an explanation for the selective advantage of the first organism with a DNA genome, becoming resistant to cellular devices against viral RNA genomes. In this hypothesis, note that all current cellular living beings would thus be, at least in part, descendants of several DNA viruses that took control of RNA cells (figure 4.8).

This hypothesis of a viral origin of DNA has had more success than that of the thermoreduction hypothesis. The former has even been the subject of two articles written by science writers in *Science* and *Nature*, which were based on a paper I had published in the French journal *Biochimie*. This was not common. The paper in *Nature*, by John Whitfield, was entitled "Base Invaders: Could Viruses Have Invented DNA as a Way to Sneak into Cells?" whereas the *Science* article, by Karl Zimmer, was titled "Did DNA Come from Viruses?"[61] Of course, the "out of viruses" hypothesis was not universally accepted. For some colleagues, attributing such an important role to viruses in the history of life was too iconoclastic. How could the "nonliving" biological entities invent new proteins to replicate their DNA? Viruses had the bad reputation of being pickpockets, capable only of stealing from infected cells the proteins they needed.[62] It was in response to these sometimes extreme reactions that I was led to reflect again on the profound nature of viruses and to propose the concept of virocells that I presented in chapter 2. If one confuses the virus and the viral particle (the virion), one cannot understand the creative nature of viruses. By contrast, if one focuses on the infected cell, the virocell, one understands well how new genes, and thus new proteins, can appear in viruses under the effect of different types of mutations during the replication of their genomes, just as they can appear in cells coding for ribosomes during the replication of their chromosomes.[63] If the "out of viruses" hypothesis is correct, then the first enzymes responsible for the transformation of RNA into DNA originated in virocells.

The "out of viruses" hypothesis implies that viruses are very ancient, since they would have already existed at the time of the transition of RNA to DNA. Indeed, viruses are most likely older than LUCA itself, since they were probably already diversified at the time of LUCA. How do we know that? This directs us to one of the major discoveries in biology at the turn of the century. It was a discovery that, for me, rivals that of the discovery of archaea. To reveal it we will begin by taking a detour to the far north, where in winter the Baltic Sea is transformed into a field of ice around the Scandinavian coast.

March 2, 2006: Helsinki

I'm walking on ice. It is a surprising sensation—and a bit worrisome for someone who barely knows how to swim. In winter it is possible to walk from the port of Helsinki to a small island a stone's throw from the Finnish capital. We pass a sailor who is repainting the hull of his boat, his two feet firmly planted on the frozen ground. We're very far from hell and hyperthermophiles. I came to this northern capital to meet Dennis Bamford (figure 4.9), one of the greatest living virologists. It is he and his colleagues who obtained the first proofs of the ancientness of viruses. He invited me to present my theories on the viral origin of DNA at a colloquium on the origins of life that he organized for his students. I was accompanied by Antonio Lazcano, my Mexican friend, a great specialist on the subject. Bamford worked for years on the PRD1 virus, which infects the bacterium *Escherichia coli*. Unlike most viruses infecting bacteria, PRD1 (or at least its virions) has neither a head nor a tail. It has an icosahedral protein capsid that encloses a lipid vesicle, itself enclosing the viral genome (a linear double-stranded DNA molecule). At the end of the 1990s, Bamford began collaborating with crystallographers Stacy Benson and Roger Burnett from the Wistar Institute in Philadelphia to obtain high-resolution structures of the major protein of the capsid of PRD1. Burnett had been studying the structure of adenovirus, a human pathogen, for a long time. He was tiring of this structure and was looking for something different, like the capsid protein of a virus infecting a bacterium. So he was delighted when Bamford suggested that he study PRD1, but was very surprised by the results.

Bamford remembers well a late-evening phone call from Burnett telling him the results: "Can you imagine, Dennis? The PRD1 capsid protein fold resembles that of adenovirus, like two drops of water."[64] This was a surprising result because their amino acid sequences do not have any detectable

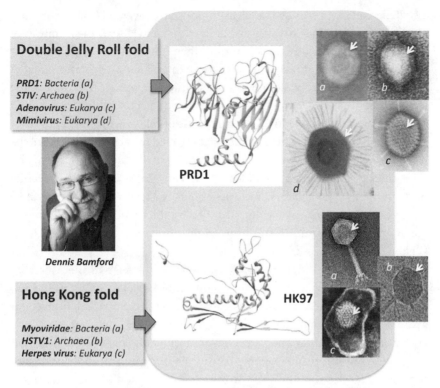

Double Jelly Roll fold

PRD1: *Bacteria (a)*
STIV: *Archaea (b)*
Adenovirus: *Eukarya (c)*
Mimivirus: *Eukarya (d)*

PRD1

Dennis Bamford

Hong Kong fold

Myoviridae: *Bacteria (a)*
HSTV1: *Archaea (b)*
Herpes virus: *Eukarya (c)*

HK97

FIGURE 4.9. The two major universal lineages of DNA viruses. The central images represent models of the major capsid proteins (MCP) of viruses from the PRD1/Adenovirus lineage (*top*) and from the HK lineage (*bottom*). The polypeptide chain is folded in the so-called double jelly-roll and Hong Kong folds, respectively (courtesy of Mart Krupovic, Institut Pasteur). A few names of viruses from each lineage infecting members from the three domains of life are indicated. Letters in brackets refer to the electron microscopic images showing the capsid (*white arrows*) of these viruses. The capsid of Myoviridae and HSTV1 correspond to the head of these tailed viruses. The herpes virion contains an envelope surrounding the capsid.

similarities; but despite these differences, the protein chains fold in the same way in three-dimensional space. It is a phenomenon that one often observes when two proteins that have the same origin have diverged for a long time: their resemblance in terms of sequences of amino acids disappears as a result of the accumulation of neutral mutations that do not, however, challenge the ability of the polypeptide chain to fold correctly in space. In that case, only the comparison of the structure of the two proteins in three dimensions can reveal their common origin.

Burnett was at first disappointed at not escaping the adenovirus, but he quickly realized that he had made a major discovery with Benson and Bamford. The capsid proteins of PRD1 and of the adenovirus are so similar that they could not have appeared twice independently. One cannot escape a striking conclusion: these two viruses must have diverged from a common viral ancestor that must have existed at the time of LUCA. These results were completely unexpected because, up until then, most biologists thought that the viruses infecting prokaryotes and those infecting eukaryotes originated independently. According to the so-called escape hypothesis that ruled at the time,[65] the viruses infecting bacteria and those infecting eukaryotes would have appeared independently from fragments of cellular chromosomes (pro-karyotic or eukaryotic), which would have escaped to become parasites of the cells from which they were issued. The discovery of Bamford, Burnett, and their colleagues challenged this idea by showing that some viruses infecting bacteria share homologous characteristics (derived from a common ancestor) with viruses infecting eukaryotes. This was another nail in the coffin of the eukaryotic/prokaryotic vision of the living world, which had already been shaken by Woese's work.

Subsequently, in a hot spring of Yellowstone, the team of the American scientist Marc Young discovered a virus infecting *Sulfolobus* called STIV (*Sulfolobus* Turreted Icosahedral Virus), whose capsid, decorated by turrets, is also formed by a major capsid protein homologous to those of PRD1 and adenovirus. It was Zillig who had given Young and his colleagues all the tools they needed to succeed in isolating the viruses of hyperthermophile archaea, during the last summer he spent in Bozeman as a guest scientist at the age of seventy-eight. The discovery of an archaeal virus sharing the same type of capsid protein with PRD1 and adenovirus confirmed that similar viruses probably existed at the time of LUCA.

Viral Lineages: A Major Advance in Virus Classification

The folds of the PRD1, adenovirus, and STIV major capsid proteins have the nice name "double jelly-roll fold," as they are formed of two sub domains, each of them having the shape of an English jelly-roll cake. All viruses charac-terized by the presence of this fold in their capsid are now considered to form a viral "evolutionary lineage" named the PRD1/Adenovirus lineage.[66]

A second very ancient viral lineage was identified in the 2000s. Surpris-ingly, this one groups the herpes viruses, specific to vertebrates and well known by most of us, with bacterial and archaeal viruses producing head and tailed virions.[67] The major capsid protein from the heads of these viruses shares with the major capsid protein of herpes viruses a similar fold (named the "Hong Kong" fold because it was first identified in a head and tailed virus isolated in Hong Kong), the bacteriophage HK97. The Hong Kong fold and the double jelly-roll fold are completely unrelated (figure 4.9).[68]

This shows that viruses of these two lineages do not descend from the same common ancestor, but originated independently.[69] Thus there exists no equivalent of LUCA in viruses.[70]

RNA viruses infecting bacteria and those infecting eukaryotes form a third major lineage of viruses since they share a homologous capsid protein characterized by a fold corresponding to a single jelly-roll domain.[71] RNA viruses were thus also probably around at the time of LUCA. The first viruses were most likely RNA viruses infecting RNA cells, in agreement with the "out of viruses" scenario for the origin of DNA, which implies that the first DNA viruses originated from RNA viruses. Several scientists indeed think that retroviruses, with RNA genomes but with DNA intermediate in their reproductive cycle, reflect the ancient RNA to DNA transition in the history of life.

When I mention that viruses are more ancient than LUCA (the logical conclusion of the work of Bamford and a few other virologists), some of my colleagues believe I am reverting to the old idea that viruses appeared before cells. These colleagues apparently confuse cells and modern cells (LUCA and its descendants). I think, of course, that viruses appeared after cells, since viruses always needed to parasite a cell to synthesize their own proteins using cellular ribosomes. Viruses probably appeared relatively late in the history of life (compared to cells), if we define viruses as organisms producing virions and if we consider that a virion should contain at least one protein coded by the viral genome.[72] In that case, the ancestors of today's viruses necessarily appeared after the emergence of RNA cells that were already very evolved, capable of producing ribosomes. Of course, it is likely that some kinds of protoviruses (molecular parasites) made only of RNA also existed from the very beginning of the RNA world, since life and parasitism have always evolved hand in hand. However, we will never know what these protoviruses looked like.

We saw in chapters 1 and 2 that the viruses of microbes from hell are very different from those that infect bacteria and eukaryotes; some of them have capsid protein with unique folds. However, most of these viruses must descend from ancestral viruses that also existed at the time of LUCA. Many different viral lineages were thus already present at that time. We can imagine that, from the ancestral virosphere, three distinct groupings of viruses, each of them containing different viral lineages, were selected at the time of the formation of the three domains: archaea, bacteria, and eukaryotes. These three viral groupings then evolved in parallel with the cells of their respective domains, which would explain why viruses infecting archaea, bacteria, or eukaryotes today seem very different at first glance. However, some viral lineages must have been represented in the three initial groupings, in particular those corresponding to the viruses with double jelly-roll or Hong Kong folds, which would explain why they are represented today in the three domains of life.[73] Other viral lineages may have been present in only one or two viral groupings accompanying the newly emerging cellular lineages. This would explain, for instance, why RNA viruses seem to be present only in bacteria and eukaryotes but not in archaea, or why other viral lineages seems to be specific to either archaea, bacteria, or eukarya.

The classification of viruses into viral lineages based on the structure of their capsid proteins was proposed by Bamford in 2003.[74] It was a great advance in viral taxonomy. First, it is a natural classification based on evolutionary relationships, as in the case of the natural classification proposed by Woese for cellular organisms. The capsid protein is thus for viruses the equivalent of ribosomal RNA for cells, and viral lineages can be viewed as the equivalent for viruses of what domains are for cells. Focusing on the virion makes sense because the virion corresponds to the hallmark of the virus. To be convinced of that point, one need only realize that the production of virions is the only feature that distinguishes viruses from other mobile elements, such as plasmids.[75]

These last few years, Didier Raoult and his colleagues, who discovered the giant Mimivirus (a real monster from a viral point of view), have suggested including viruses with large DNA genomes in a fourth domain of life.[76] I don't agree with them because these giant viruses are indeed viruses. They don't code for ribosomes, and most of them possess capsid proteins of the double jelly-roll fold type. Thus they are part of the PRD1/adenovirus lineage. I recently wrote an article with David Prangishvili and Mart Krupovic to clarify our point of view: to define a domain of life, one must have ribosomes, which was Woese's initial idea. In the case of viruses, one should speak of viral lineages (as defined by Bamford) and base one's definition on the structure of the virion.[77]

Where Are Viruses in the Universal Tree of Life?

If viruses cannot be grouped into domains, where can they be placed on the universal tree of life? All modern cells, with few exceptions, are infected with a large number of different viruses. There is no reason to think that it hasn't always been this way, at least since the appearance of viruses on Earth. If we look at the universal tree of life and raise the question "Where are the viruses?" we must then imagine that they are here, there, and everywhere. Viral lineages are interlaced around the trunk of the tree, the branches, and the leaves like ivy in a virgin forest, but they don't suffocate the tree. On the contrary, thanks to their creativity, by bringing new genes and forcing cells to resist them, viruses are a major factor in evolution. They have probably played a determining role in several key stages of the history of life. For example, I think, along with Prangishvili, that the "invention" of the peptidoglycan by bacteria could have been, above all, a means to prevent viruses from infecting them. By cloning and expressing in the bacterium *Escherichia coli* the gene coding for the protein that forms the pyramids produced by the archaeal virus SIRV2 to expulse their virions outside the cell (remember chapter 2), Tessa Quax, one of Prangishvili's students, has shown that this bacterium indeed produces pyramids at the level of its membranes, but these pyramids are not able to pierce the peptidoglycan.[78] The existence of peptidoglycan in bacteria could explain why the diversity of viruses that infect archaea is much greater than that of viruses that infect bacteria, 95% of which are head-tail viruses of the HK97 lineage. The viruses infecting bacteria are doubtless those that have succeeded in circumventing the defense mechanism corresponding to peptidoglycan. I also think that the nucleus of eukaryotic cells might have appeared at first as an antiviral defense mechanism. To conclude, the arms race implied by the competition between viruses and cells has surely played a primary role in the history of life on our planet.[79]

In ending this chapter, let's go back to Woese, who several times mentioned that the aim of his work was to fulfill Darwin's great dream. This dream was described in an 1857 letter to Thomas Huxley in which Darwin wrote, "The time will come I believe, though I shall not live to see it, when we shall have very fairly true genealogical trees of each great kingdom of nature." We are rather close to this point now with Woese's tree of three domains infected by numerous ancient viral lineages. However, following the genomic revolution, controversies are still raging, especially concerning the location of eukaryotes (humans) on the tree. This will be one of the main focuses of the next chapter. Are microbes from hell our grandfathers or our great-uncles? Unfortunately, Carl Woese will not be there to see the end of the story.

January 12, 2013: The Death of Ourasi

Seated in front of my TV, I am uneasy; the evening news announcer has just reported the death of Ourasi, the most popular French horse of all time. With a solemn air the announcer recalls the prestigious career of this superb trotter, born April 7, 1980, a four-time winner of the America's Cup. Two weeks earlier I had learned, through an e-mail from a colleague, of Carl Woese's death on December 30, 2012. At the time, Woese, one of the greatest intellects of the twentieth century, didn't merit a line in the newspapers or a word in the televised newscasts. This demonstrates well the place of science in the collective imagination and in the minds of modern media editors. I like horses, but still . . . I picked up my pen and sent an article to the newspaper *Libération*. The text wasn't published, but it was posted on *Libération's* website.[80] I have reproduced below part of the article, in homage to a great man who changed the course of history.[81]

> I learned that Ourasi had died on the 8:00 p.m. news, ten minutes of prime time, at least it seemed that long to me. It seems everyone knew Ourasi, a mythical horse that won the America's Cup four times. I'm sorry, but I didn't know him. Like you, I like horses, so I was somewhat saddened as I watched the news. Not long before, the world lost one of its greatest scientists, the American Carl Woese. Some say he was the greatest biologist of the twentieth century. I didn't expect to hear of his death on prime time, of course, but at least to read an article in the daily press. But no, there was nothing. The death of this giant of science remained unnoticed in our country. Brief mentions in *The New York Times* and a few American journals, a few blogs on the Internet, just enough to satisfy specialists who already knew him. This comparison with the media treatment given to Ourasi is not surprising, and that's what makes me so indignant. Carl Woese revolutionized our conception of the world. He was a modern Copernicus. Before Copernicus, Earth reigned in majesty at the center of the solar system. Before Carl Woese, life on Earth was classified into five kingdoms: animals, plants, mushrooms, protists, and bacteria. . . . Biodiversity seemed above all represented by macrobes, organisms visible to the naked eye, our faithful dogs, and our green plants. Thanks to the tools invented by Carl Woese, we know today that 95 to 99% of the biodiversity on Earth is composed of microbes and their viruses. Carl Woese and his disciples succeeded in positioning all these microorganisms into a universal tree of life, going back in the history of life up to our last universal common ancestor. Among scientists, Carl Woese is above all known for his discovery, in 1977, of a new world, that of archaea, microbes that had stayed hidden from view for centuries, confused with bacteria. Since then we know that the living world is divided into three large evolutive domains: archaea, bacteria, and eukaryotes, each of them associated with multiple viral lineages. But have you ever heard of archaea? How many announcers on the 8:00 p.m. news, how many politicians have heard of archaea? At least the king of Sweden, a descendant of Bernadotte, has heard of them. In 2003 he awarded Carl Woese the prestigious Crafoord Prize, awarded every three years to a biologist, for his discovery. . . .

The Universal Tree of Life: Are Microbes from Hell Our Ancestors?

The Saga of Genomes

Genomics, the science of genomes, represents one of the greatest scientific advances of the last century; its effects have revolutionized all of biology. It was born of the discovery of the structure of DNA in 1953 and many technological developments resulting from the biochemical study of enzymes involved in its replication, in particular DNA polymerases.[1] Not long after an enzymatic DNA sequencing method was perfected by Sanger in 1977, a small group of scientists came up with an idea that, at the time, seemed a bit crazy: to decode the entire human genome. They would attempt to determine the chaining order of the four DNA bases—A, T, G, and C—over all our chromosomes, amounting to more than 3.4 billion letters. This dream became reality ten years later, owing, among other things, to the perfection and automation of the polymerase chain reaction (PCR) method (made possible by the DNA polymerase of microbes from hell).[2] The scientists now needed only to find funding and to convince labs to get to work. This project, first known as HUGO (HUman Genome Organization), would have major repercussions in all realms of biology, including that of microbes from hell. Let's go visit Carl Woese one more time, at the beginning of the 1990s, to understand how a very informal organization was born, one that inspired many scientists, including me: ARGO (ARchaeal Genome Organization). We will see how history then sped up, in a way no one could ever have imagined.

March 1993: Munich—Carl Woese's Prophesy

In March 1993 a colloquium was held in Munich to mark the official retirement of Wolfram Zillig—who, fortunately, remained very active for another ten years. At that time, the complete sequencing of the human genome still seemed a very distant possibility, but while waiting, many labs had started on much smaller genomes belonging to organisms that were considered models—that is, studied by a large number of scientists (the *Escherichia coli* bacterium, baker's yeast, *Saccharomyces cerevisiae*, and some others). The initial results were expected at the end of the millennium. It was high time for the community of scientists studying archaea to take their turn and enter this genome race by attacking the genome of an archaeon. To get the project under way, some participants at the Munich colloquium, including me, decided, as we've just seen, to form a small, informal group, pompously called ARGO in reference to HUGO. A small meeting was organized in the library of the Max Planck Institute by Roger Garrett, a pioneer of research in the molecular biology of archaea, to reach an agreement on the name of the archaeon that would be chosen to spearhead this project. Around the small table, each person offered the name of their favorite organism, always with good reasons to defend it. Woese was at this meeting. As usual, he didn't say very much. Then he indicated that he would like to speak. Everyone was immediately silent. The oracle was going to speak; we were going to learn the name of his favorite archaeon.

"We're going to have to sequence at least five [genomes]," he said before falling silent again. Everyone was instantly uncomfortable. Had Woese lost his mind? At the time, sequencing a genome seemed to be a huge challenge, and many biologists thought that once that mountain had been climbed it would be wise to stay at the top, at least for a while, before beginning that titan undertaking again. The future would prove once again that Woese saw farther than the other people there. At the end of the 1990s, five archaea genomes had been entirely sequenced. Today, we have more than two hundred!

Of all the participants at the ARGO meeting, the first to get started was of course Woese himself. To go straight to his goal, he knocked on the door of the most effective sequencer at the time, Craig Venter. Venter had caught the competition napping when, at the beginning of the 1990s, he created a private institute devoted entirely to the sequencing of genomes: TIGR (The Institute for Genomic Research). In 1995 the scientists at this institute published the first complete sequence of the genome of a cellular organism, the bacterium *Haemophilus influenza*.[3] Thanks to the collaboration between Woese and Venter, the third genome of a cellular organism ever sequenced—published in 1996—was that of an archaeon and, what is more, of a microbe from hell, *Methanocaldococcus jannashii*.[4] Not long afterward, the first genome of a eukaryotic organism, the yeast *Saccharomyces cerevisiae*, was published by a European consortium. It is the comparative analysis of these first three genomes that confirmed the division of the living world into three domains, which in large part explains the awarding of the Crafoord Prize to Woese in 2003. It became apparent that archaea differed from bacteria and eukaryotes not simply by their 16S ribosomal RNA and their lipids but also by the great majority of their proteins.[5] As we saw in chapter 2, for most proteins coded by the genome of *Methanocaldococcus jannashii*, we indeed found an archaea-type version that was distinct from bacterial and eukaryotic versions of the same proteins. This genome also coded for a great deal of proteins that didn't resemble any other protein recorded in any database up to then.

The other scientists at the ARGO meeting ultimately obtained "their" genomes as well. In particular, Garrett was able to organize a consortium among Canadian, French, and Danish scientists to sequence the genome of *Sulfolobus solfataricus*. Yvan Zivanovic, a scientist from our BMGE team at Orsay, made a decisive contribution to this project in collaboration with Fabrice Confalonieri, who worked at the time with Michel Duguet at our Institut.[6] I had the good fortune to collaborate with Génoscope, a French lab devoted entirely to sequencing genomes, set up in Évry in 1998 and directed by Jean Weissenbach. At our suggestion, Weissenbach chose as the first study model to train his team in genome sequencing a microbe from hell, *Pyrococcus abyssi* (only two million base pairs), which had been isolated ten years earlier by Prieur's team in Brest. I had the great pleasure of annotating the genome of *P. abyssi* (that is, identifying the genes present on this genome and the proteins coded by these genes) in collaboration with George Cohen, a pioneer in molecular biology in France, who had worked with François Jacob and Jacques Monod at the Institut Pasteur.[7]

The sequencing of the *P. abyssi* genome enabled us, through bioinformatics, to identify the starting point of the replication of DNA on the chromosome of this organism, as well as the protein responsible for the beginning of the replication, a protein from the same family as those that initiate the replication of chromosomes in eukaryotes.[8] While carrying out an experiment to confirm this result, a Finnish scientist doing a postdoctoral fellowship with our team at Orsay, Hannu Myllykallio, noticed that our pet archaea must not have possessed DNA. In fact, it lacked an enzyme, thymidylate synthase, called ThyA, enabling it to go from the letter U (of the RNA) to the letter T (of the DNA). This observation led him to the discovery of a new enzyme capable of catalyzing this reaction: ThyX.[9] The crystallization of the protein ThyX from the hyperthermophilic bacterium *Thermotoga maritima* by the Joint Center for Structural Genomics[10] enabled us to compare its structure in space to that of ThyA and to see that these two proteins do not resemble each other and thus are not homologues, which means they do not have the same origin. This observation taught us that going from the letter U to the letter

T in the DNA must have occurred twice independently during evolution.[11] Note that many viruses code for the proteins ThyA and ThyX, as well as for ribonucleotides reductases (RNR)—enzymes that transform ribonucleotides into deoxyribonucleotides. The presence of genes coding for these enzymes in viral genomes is for me an additional argument in favor of the hypothesis of a viral origin of DNA. Indeed, this shows that the viral world provides all the proteins necessary to go from the RNA world to the DNA world: those that enable the transformation of substrates of RNA into substrates of DNA (ThyA, ThyX, and RNR), those capable of transcribing RNA into DNA (these are the inverse transcriptase of retroviruses), and those that enable the replication of viral DNA (DNA polymerases, helicases, or even topoisomerases).

The discovery of ThyX was not just of academic interest. Myllykallio had in fact observed that the genes coding for this protein were present in many pathogenic bacteria that didn't have ThyA, the protein present in humans. The protein ThyX is thus a good target for research into new antibiotics against those bacteria, an unexpected path leading microbes from hell to a possible medical application.[12] Myllykallio's work enabled us to publish two papers in *Science* in two years, 2000 and 2002, which probably explains why I was recruited by the Institut Pasteur in 2003. Thank you, Hannu and genomics.

At the beginning of the 2000s, history accelerated even more. The human genome was sequenced in 2001, in tandem by a public international consortium and a private American company, Celera, created by the indomitable Craig Venter. Ten years earlier, everyone thought that the adventure of sequencing stopped there. On the contrary, the methodology became routine and the equipment available, and sequencing was to accelerate exponentially. By August 2015, when I was writing this book, more than five thousand genomes had been sequenced, including a large number of bacterial genomes, but also a great many eukaryote genomes and, as we've seen, more than two hundred genomes of archaea.

Today, many labs are systematically sequencing a large number of genomes of very similar organisms to better understand their evolutive processes. Within the framework of my European project, EVOMOBIL, in collaboration with Génoscope, we have planned to sequence the genomes of a hundred microbes from hell belonging to the order of *Thermococcales* to observe how the mobile elements that infect these organisms, plasmids and viruses, influence the evolution of the genomes. Jacques Oberto, a scientist at CNRS, an expert in genetics and bioinformatics, will pilot this work.[13] Oberto took over the direction of our team at Orsay after I retired from the university. Thanks to the funding from the European project, I was able to be recruited as a professor at the Institut Pasteur (where I worked for free for ten years) to pursue with Prangishvili my work on archaea and viruses of microbes from hell.

In this brief description of the history of genomics, I've used the word "bioinformatics" twice. Bioinformaticians have played a key role in the analysis of genomic sequences. With their help, we first had to identify all the proteins coded by these genomes, some hundreds of thousands, and classify them into families so we could compare the proteins from the same family among different species. This work enabled us to show that most well-preserved proteins in general existed in three versions—one for each of the three domains—and to identify the transfers of genes between the organisms, which I discussed in chapter 2 (figure 2.6). The existence of gene transfers had been known for a long time, but the analysis of the genomes showed that they occurred more frequently than we had imagined up to then. For example, we noticed that the archaea that lived in acid environments, such as *Thermoplasma acidophilum* and *Sulfolobus acidocaldarius*, had exchanged many genes, even though they were very far from each other on the evolutive level, since the first is a Euryarchaeota whereas the second is a Crenarchaeota. As I briefly noted in chapter 2, the same is true of bacteria and hyperthermophile archaea that live in the same hot springs. As a general rule, cohabitation favors the exchange of genes, a bit like the cohabitation of two groups of humans speaking different languages favors the exchange of words between the languages.

Toward a Robot Portrait of LUCA

The comparison of proteins coded by the genomes of archaea, bacteria, and eukaryotes enabled us to identify the proteins present in all organisms[14] in the three domains of the living world. These universal proteins were probably already present in LUCA. Their identification has enabled the establishment of a first robot portrait of LUCA. There aren't many of these proteins (from sixty to one hundred, depending on the computational method used to detect them).[15] Of course, LUCA must have had more than sixty genes, perhaps a few hundred, to be able to ensure the existence of a metabolism capable of providing the energy necessary for the synthesis of all its cellular constituents. Many proteins present in LUCA most likely didn't leave any descendants. Some of them must have been lost in one or the other of the three domains. Others could have been replaced by proteins with the same function but of different origin; these are called analogous proteins.

In the small group of universal proteins we find mainly those that intervene in the synthesis of proteins (ribosomal proteins, elongation factors, enzymes fixing amino acids on transfer RNA) and RNA polymerase that transcribes DNA into RNA. By contrast, as we saw in the previous chapter, we don't find in the list of truly universal proteins those involved in the replication of DNA, since those are not homologous among either bacteria or archaea/eukaryotes. We have seen that this observation led to the idea of an RNA-LUCA. But in that case, why, if LUCA didn't have DNA, do we find in the list of universal proteins RNA polymerase, which transcribes DNA into RNA? I have an explanation (hypothetical, of course). It happens that in eukaryotes the RNA polymerases used to manufacture messenger RNA from the DNA of chromosomes are also used by certain RNA viruses to replicate their genomes.[16] These enzymes can thus read at the same time the genetic messages of DNA and RNA. In this case, the ancestor of our RNA polymerase could have been used both to transcribe and to replicate the RNA genome of LUCA.

Around half of the universal proteins, thirty-four, are ribosomal proteins. If we root the universal tree in the bacterial branch, as suggested in chapter 4, the other proteins found in modern ribosomes were probably added after the divergence of lineages leading to bacteria and to archaea/eukaryotes. This means that the ribosome of LUCA contained only thirty-four proteins[17] (figure 5.1a).

It was thus smaller and simpler than current ribosomes. This conclusion aligns with Woese and Fox's initial idea of a LUCA whose molecular biology was more primitive than that of modern organisms—it was a progenote.[18] Woese and Fox even thought that LUCA was incapable, with its miniribosome, of manufacturing proteins in as reliable a way as is done today, which led to the manufacture of proteins whose sequence of amino acids always reflected that of the gene. In their opinion, the proteins manufactured by LUCA could have been a mixture of slightly different polypeptides, only part of them able to function effectively.

A research project carried out by our BMGE team at Orsay leads me to believe that this is not the case. When I told you about our celebratory meal at La Crémaillère in Bourg La Reine in 2014, I forgot to introduce two of the guests: Tamara Basta, assistant professor at Université Paris-Sud in Orsay, and her doctoral student, Ludovic Perrochia. Ludovic had just defended his thesis a few weeks earlier. For three years he had studied two universal proteins whose roles had remained unknown for a long time. At the time he began his work, an American team, led by a French scientist, Valérie Crecy-Lagarde, had just shown that these two proteins, originally named Kae1 and Sua5 in archaea,[19] intervene together in the catalysis of a universal chemical modification that affects all transfer RNA that read codons (triplet) beginning with

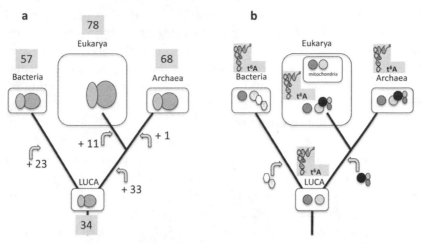

FIGURE 5.1. Rooting the tree of life in the bacterial branch using comparative molecular biology. *A*, evolution of the ribosome protein content. The numbers of ribosomal proteins in modern ribosomes and in LUCA are in gray squares. When the tree is rooted in the bacterial branch, evolution can simply be explained by the addition of specific ribosomal proteins at each step of evolution. *B*, evolution of the proteins involved in the synthesis of the tRNA modification base t6A. Two homologous universal proteins (*dark and light gray circles*) and two (bacteria) or three (archaea/eukarya) accessory proteins are required to perform this reaction in modern organisms. The two accessory proteins are sufficient to perform the reaction in mitochondria and were probably also sufficient at the time of LUCA. One set of accessory proteins was added in the branch, leading to bacteria, whereas the other was added in the branch common to archaea and eukarya.

the letter A.[20] This modification, called t6A, corresponds to the addition of an amino acid, threonine, and a carbamoyl grouping on an adenine of a transfer RNA close to an anticodon. It is essential to stabilize the correct interaction among codons beginning with the letter A and the corresponding anticodons because, as you might remember from chapter 3, the AU base pair involves only two hydrogen bonds (figure 3.2), which is not enough to ensure the specific interaction of a codon/anticodon (e.g., ACU/UGA) starting with an AU base pair.

In archaea, bacteria, and eukaryotes, Ludovic, Tamara, and two other research groups in the United States and Canada have shown that the two universal proteins, Sua5 and Kae1, need several accompanying proteins to catalyze the formation of t6A in archaea and eukarya. As in the case of proteins that replicate DNA, the accessory proteins of bacteria are not homologous to those of archaea and eukaryotes (figure 5.1). Thus they were probably not present in LUCA. At first sight, this conclusion seemed to agree with the idea of a LUCA incapable of manufacturing proteins in a very reliable way, since it could not catalyze a modification of transfer RNA essential to the correct reading of the genetic code. However, I had trouble imagining such a deficient

LUCA. It was, all the same, at least capable of manufacturing the hundred universal proteins that kept enough sequencing similarity between the versions of the three domains to be recognized as homologous after more than three billion years. I was reassured when Ludovic and Tamara, in collaboration with Valérie and in parallel with the Canadian group, showed that the two universal proteins present in the mitochondria were sufficient to catalyze the formation of t6A in the transfer RNA of these organelles.[21] In fact, during evolution the mitochondria lost the accessory proteins of their bacterial ancestor. We then thought that the two universal proteins present in LUCA were probably sufficient at the time to catalyze the formation of t6A, as they have once again become capable of doing in mitochondria (figure 5.1b). This observation suggests that LUCA, even if it were simpler than modern cells, was already capable of faithfully translating the RNA message and of manufacturing good-quality proteins since it took the trouble to modify its transfer RNA to be able to correctly read the codons beginning with the letter A.

In chapter 4, we saw that Koonin and Mushegian were the first in 1996 to notice the great DNA replication divide, suggesting that LUCA had an RNA genome.[22] Ten years later, Koonin and his colleagues revealed another great divide from comparative genomics, this time in the mechanism of ATP production, discussed in chapter 3.[23] We have seen that the ATP synthases that manufacture ATP from the energy provided by the gradients of proton or sodium are formed by the association of several subunits, one of which could be called the rotating subunit (figure 3.8). Remember that this subunit plays a key role in the story since it can spin on itself under the effect of an electrical current, which enables the conversion of electrical energy into chemical energy to manufacture ATP. Strikingly, all the subunits of ATP synthases are homologous in bacteria and archaea/eukaryotes, except one: the rotating subunit. To explain this unexpected observation, Koonin and Armen Mulkidjanian advanced the idea that the ancestor of the modern ATP synthases, which was present in LUCA, did not have a rotating subunit, and thus could not manufacture ATP. In their opinion, this protein instead consumed ATP to expulse the sodium ions toward the external environment, playing the role of a reverse sodium pump. Subsequently, two distinct types of rotating subunit would have been invented independently, one in the branch leading to bacteria, the other in the branch leading to archaea and eukaryotes; this would have enabled modern organisms to have a new, very effective mechanism to synthesize their ATP.

If this hypothesis is correct, it means that LUCA manufactured its ATP through processes of fermentation, like some rare eukaryotes that have lost mitochondria in the course of evolution (this is the case for the intestinal parasite Giardia lamblia), or like mutant yeasts devoid of mitochondria. This hypothesis once again goes very well with the idea of a LUCA that was simpler than modern cells. It also suggests an interesting connection between LUCA and humans—eukaryotes. In fact, eukaryotes have an enzyme that greatly resembles the archaeal membrane ATP synthase (with its rotating subunit), but this protein functions in reverse by using the energy of ATP to transport ions or protons or both through the membranes of our vacuoles. As a result, eukaryotes don't use their archaeal-like ATP synthases to manufacture their ATP; instead, they depend on their bacterial-like ATP synthases, present in their mitochondria, for ATP synthesis. Protoeukaryotes (Woese's urkaryotes) thus must have also used fermentation to manufacture their ATP. This question takes us to the problem we will face at the end of this chapter: the origin of eukaryotes and our kinship relationships with archaea.

These results enable us to define a "minimal LUCA": a cellular organism whose genome was perhaps still made up of RNA and that synthesized its proteins through a simpler mechanism, but one that was similar to that used by modern cells by using the same genetic code as that which is currently in use. At some point, some scientists, including Koonin, even imagined that this rather primitive LUCA was not yet cellular. They imagined that it was still living in the mineral cages of a hydrothermal chimney and that two different types of lipids and membranes emerged only after the divergence of the archaeal and bacterial lineages from LUCA, explaining the great lipid divide.[24] However, the analysis of universal proteins definitely shows that LUCA already possessed a cytoplasmic membrane.[25] We have just seen that all subunits of the membrane bond ATP synthases, except the rotating subunit, are homologous in the three domains of life, suggesting that LUCA contained a membrane bond sodium pump. The set of universal proteins also includes membrane proteins required for the secretion or the entry of macromolecules through the membrane. Even more striking, the universal protein set includes a multisubunits proteins complex, called the signal recognition particle (SRP), which is required for the insertion of membrane proteins in the lipid layer of the cytoplasmic membrane. The SRP interacts with ribosomes in the process of manufacturing a membrane protein (signaled by a hydrophobic tail) and directs this protein toward the membrane to facilitate its insertion.

We thus now have a rather reliable robot portrait of LUCA. Some people occasionally wonder if LUCA really existed or if it is a kind of virtual organism or a collection of cells exchanging genes—the communal LUCA, an idea first suggested by Woese.[26] In fact, we can say that LUCA really did exist. LUCA is the logical consequence of the binary mechanism of cell division (one cell produces two cells): the coalescence point of all modern lineages going back in time for each of them, from daughter cells to mother cells and so on. Similarly, the true existence of the African Eve, the last ancestor common to all women living today, is the logical consequence of the direct transmission of mitochondrial DNA, undisturbed by recombination, from mother to daughter. The comparison between LUCA and the African Eve also helps us to realize that LUCA was not a solitary type. It must have shared the planet with a multitude of other organisms, its contemporaries, more than three billion years ago. Similarly, the African Eve certainly didn't live alone in her village 150,000 years ago. Among all their contemporaries, LUCA and the African Eve were simply the only ones that left descendants on our planet. The existence of LUCA's contemporaries should also be taken into account in comparative genomic analyses. In addition to universal proteins inherited from LUCA, modern organisms might have inherited from LUCA's contemporaries some genes that were transferred a long time ago from those organisms to ancestors of modern organisms, before their lineages became extinct.

The Root of the Universal Tree Seems Confirmed

We saw in chapter 4 that phylogenetic analyses have suggested rooting the universal tree of life between bacteria and a common ancestor of archaea and bacteria, but also that I was for a long time skeptical of this. However, I have finally been convinced by all the data from comparative genomic analyses that I have just discussed. A bacterial rooting better explains the great replication and ATP synthase divides, as well as many other differences between the molecular biology of bacteria and of archaea and eukarya. Otherwise, one has to assume that all features common to archaea and eukarya were already present at the time of LUCA, and that all of them were replaced independently in bacteria. For example, if we refer to the ribosome protein composition, we should imagine that the thirty-three proteins common to archaea and eukarya were replaced in the bacterial lineage by twenty-three new ribosomal proteins (figure 5.1a), and that the three accessory proteins involved in the biosynthesis of t6A common to archaea and eukarya were replaced by the two nonhomologous accessory proteins specific to bacteria[27] (figure 5.1b). The same reasoning (multiple protein replacements in the bacterial branch) can be made for DNA replication proteins or for the rotary subunit of the ATP synthases. This scenario cannot be completely excluded, but it seems very unlikely. In this scenario, it is hard to see why so much drastic replacement should have occurred specifically in the bacterial branch, transforming an archaeal/eukaryal-like LUCA into a bacterium. There is no obvious selection pressure that could justify all these replacements. I definitely now prefer the idea of a simple progenote-like LUCA that was transformed independently into the three modern types of cells.

Many teams of bioinformaticians have been involved in the great comparative genomics game. One team stands out from the others by the quantity and importance of its published work. This is Koonin's team, working at the National Center for Biotechnology Information (NCBI) in Bethesda, Maryland. I am particularly grateful to Koonin and his close colleagues, especially Michael Galperin, another bioinformatician of Russian origin, for having perfected (at the beginning of the 2000s) a computer program that, as we will soon see, enabled me to make a fundamental observation about our microbes from hell. What is more, Koonin is one of the rare evolutionists who places the same importance on viruses as I do. It is thus time that we get to know him better. Indeed, he enjoys good food, as many scientists do. As you must have noticed, we do like to gather for a good meal.

October 19, 2013: Paris—Oyster Bar, Boulevard Beaumarchais

The waiter brings us a platter of oysters cooled with liquid nitrogen—amazing. White smoke escapes from the platter before us and cascades onto the tablecloth. The menu is shown on an electronic tablet; in choosing the wine we can connect directly with the producer via the Internet. This is a memorable meal for four slightly tipsy scientists. We have invited Eugene Koonin, who has come to visit us at the Institut Pasteur, to join us. To my right is David Prangishvili, who has just been named an honorary member of the Georgia Academy of Sciences. David and Eugene know each other well; they have worked together to analyze the genomes of viruses from hell that infect archaea. They have shown that 90% of the proteins coded by these viruses resemble nothing, which once again demonstrates the creativity of viruses: they are at the origin of a large number of proteins that are specifically viral.[28] On my left is Mart Krupovic, the young Lithuanian scientist who has just been recruited to work with us at the Institut Pasteur. Mart completed his thesis with Dennis Bamford in Helsinki (where you can walk on the sea). Mart is enthralled by viruses, whether they infect bacteria, archaea, or eukaryotes. Now, with David Prangishvili, he is directing a doctoral student, Emmanuelle Quemin, who is studying how the virion of the SIRV (*Sulfolobus islandicus* rod-shaped virus) recognizes its victim, the archaea *Sulfolobus*, and causes its DNA to enter the cell it infects. As you may recall, it is that virus that produces pyramids to make its virions leave the infected cell. Emmanuelle has shown that the virion, a long cylindrical tube that ends with curious appendages, uses as an entry the extremity of hairs, called pili, of *Sulfolobus*.

Unbelievable, but perhaps true; it seems that the virion then moves along the pili by using its appendages as little claws to meet the cell in which it will infiltrate, we still don't know how.[29]

Today, between two oysters, Mart is discussing with Eugene his preparations for a visit to the United States, where he will work for two months with his team. During that time Mart would make a remarkable discovery, that of a new family of mobile elements that he will call Casposons.[30] According to him, these groups of genes would be the ancestors of a system for the defense of bacteria and archaea against viruses, a system with an unwieldy name: the CRISPR system. This system acts like an immune system that enables bacteria and archaea to preserve the memory of a virus they have survived. Our immune system also seems to have evolved from proteins coded by mobile elements. Mart and Eugene would soon write a paper that would appear in *Nature Reviews Microbiology* that showed the parallel between the immune system of bacteria and archaea and that of animals.[31] These systems are very different, but in both cases, they would have evolved from mobile elements. For me, these latter certainly come from ancient viruses, which once again highlights the importance of viruses in the evolution of the living world.

The discovery (in 2005) of the bacterial and archaeal antiviral CRISPR immune system today opens up onto a technological revolution—becoming famous by the name of CRISPR Cas9[32]—that will soon enable the manipulation of the human genome at will (for better or for worse, as usual). This is another striking example of the way in which purely academic research can lead to applications that no one ever imagined. That is another example to think about for politicians who have not yet understood how scientific research functions and who usually want to oversee it.

An Icelandic Scenario for the Origins of Life?

Koonin is interested in everything, which manages to irritate many colleagues who seem to find him always one step ahead. Of course, he is also interested in the origins of life. We were in complete disagreement at the time when, borrowing the ideas of Michael Russell, he supported the idea that life had appeared and developed from the beginning to the emergence of archaea and bacteria in a single hydrothermal smokestack. In the scenario he published in 2006 with a German scientist, Bill Martin, poor LUCA didn't even have a membrane. It wasn't yet a cell, but rather a group of macromolecules stuck in mineral cages that made up the hydrothermal smokestack. Viruses were already present in the form of infectious nucleic acids (I would have said plasmids) that could pass from one cage to another. Several of us pointed out that the idea of an acellular LUCA was contradicted by the existence of universal membrane proteins, which we were just talking about a few moments ago.

Since then, Koonin's thinking seems to have evolved. Again collaborating with Mulkidjanian, he recently proposed a scenario for our origins that seems much more promising to me, in a paper published in *Proceedings of the National Academy of Sciences*.[33] Mulkidjanian had again raised an old question that had for a long time remained unanswered: Why is life dependent on potassium, whereas the biotopes where it develops today are very poor in that element? We saw in chapter 3 that all living beings are characterized by very high intracellular concentrations of potassium, whence the idea that LUCA, too, must have been rich in potassium. It seems that this ion in particular is indispensable for the functioning of the active site of the ribosome, which is made up of RNA, for the formation of the peptide bond.[34] But we saw in chapter 3 that potassium has little affinity for RNA, so it takes large amounts to ensure its being fixed on the ribosome. The problem is that potassium is rare in the exterior environment, particularly the ocean. Current cells must thus expend a lot of energy to allow the little potassium present in the exterior environment to enter through the use of sophisticated proteic pumps. Why go to all this trouble? According to Mulkidjanian, it would involve an addiction that goes back to the origins of life. In his opinion, life would have appeared in a milieu that was rich in potassium. All the primal cells thus obtained this ion directly through passive diffusion through their membranes. That was good, because they could not yet have had the sophisticated proteinic pumps enabling them to capture potassium from the environment and accumulate this ion in their cytoplasm, while at the same time expelling sodium, in both cases against concentration gradients. Only later, when such pumps appeared, were cells able to begin to explore the environments that were increasingly poor in potassium and increasingly rich in sodium.

Mulkidjanian wondered if one could still find on Earth environments rich in potassium that might recall the cradle of our origins. While discussing this with geologists, he learned that such environments do indeed exist; they are steamy zones found in areas of terrestrial hydrothermal springs. Life might thus have appeared in these hot (or warm) springs before leaving to explore the rest of the planet. This does not necessarily mean that life appeared at very high temperatures, because the temperatures of some of these terrestrial springs are rather warm or are close to cold regions (or both). This is what we observe today in a country like Iceland, the land of fire and ice, where many hot springs are found alongside the country's glaciers. I remember a hike I took a few years ago in that wonderful country where I saw steam escaping from craters located right in the middle of glaciers.

For life to appear, it seems necessary to combine hot and cold. On the hot side, we think today that the RNA bases could have been formed from a small organic molecule, formamide, $HCONH_2$, which must have been abundant on primitive Earth.[35] The formation of bases from formamide necessitates a supply of energy in the form of heat or electrical discharges. Scientists have shown that the shock of the meteorites from the Late Heavy Bombardment could also have provided that energy.[36] In that case, the bombardment would have occurred well before the appearance of LUCA and thus would not have allowed the selection of microbes from hell, as has been suggested by several scientists. Still on the hot side, we think that volcanic regions could have provided the phosphate necessary for life, among other things, to form the nucleotides from the bases and from ribose. On the cold side now, scientists have shown that the polymerization of nucleotides could have occurred at temperatures lower than 0°C through a concentration of the latter in the liquid veins inside blocks of ice. Halfway between hot and cold, in an environment rich in potassium, little drops of lipids could have been formed from simple lipids and enriched in small RNA formed in the cold regions, which would have ended in the formation of the first cells (or protocells). Recent experiments by chemists in Strasbourg, France, have shown that the energy associated with the superficial tension on the internal surface of little drops may catalyze chemical reactions very efficiently.[37] The formation of vesicles could thus have initiated the reactions of condensation necessary for the beginning of a protometabolism.

There are still lots of questions remaining in this scenario, and we are quite far from LUCA here. It probably already lived in an environment that was poor in potassium and rich in sodium, because the pumps enabling the expulsion of sodium are part of the universal proteins. In my opinion, at the time of LUCA, life had already left its original cradle some time before to explore the rest of the planet.

The Reverse Gyrase: Passport to Hell?

Let's go back to our meal. Eugene Koonin has finished his seafood platter. For me, Koonin is above all the one whose team perfected a computer program called phylogenetic pattern search (PPS), which, at the beginning of the 2000s, enabled me to answer a very simple question: Are there proteins specific to microbes from hell, and if so, how many and which ones are they? If such proteins existed, they must play an essential role in the adaptation of these microorganisms to infernal conditions, and studying them would open up a new realm of research in our lab. PPS enabled the identification of proteins that were at the same time systematically present, without exception, in a particular group of organisms and systematically absent, again without exception, in another group. In the case that interested me, it was a matter of finding how many proteins are, at the same time, always coded by all the genomes of hyperthermophiles and never by the genomes of mesophiles. With a simple click of the mouse, the list of families of proteins satisfying those two criteria appeared on the computer screen. When I carried out this experiment in 2002, I expected to find a dozen or so families of proteins out of the several thousand that were already in the database. But I was surprised by the results: there was only one. To my great satisfaction, it wasn't just any old protein. It was the reverse gyrase, discovered almost twenty years earlier in the hyperthermophile archaea *Sulfolobus*, whose identification had played such an important role in the launching of my career. I published the results of this bioinformatics analysis in 2002 in the journal *Trends in Microbiology*, results that are still valid twelve years and a few thousand genomes later.[38] The gene coding for the reverse gyrase is always present, without exception, in the genomes of all microbes that can multiply at above 80°C. By contrast, it has never yet been observed in the genomes of microbes that can multiply only at temperatures below 50°C. The situation is in between for moderate thermophiles that multiply at temperatures between 50 and 80°C. Some have the reverse gyrase, but others don't.

In chapter 3 we saw that the exact role of the reverse gyrase remains mysterious. All the same, the results obtained through comparative genomics seem to indicate that this role must be very important. It suggests that you can't be a microbe from hell without the reverse gyrase. For me, these results represented another argument against the idea of a direct link between a hot origin of life and modern hyperthermophiles. Indeed, the analysis of the gene of the reverse gyrase has shown that this enzyme could not have been a primitive protein. Fabrice Confalonieri, in Michel Duguet's lab, had succeeded at the beginning of the 1990s in isolating the gene coding for the reverse gyrase of *Sulfolobus shibatae* and in deducting its sequence in amino acids. The analysis of this sequence had shown that the reverse gyrase was a composite protein, formed by the fusion of two enzymes: a DNA topoisomerase of type I, "classic," and an enzyme called helicase, which can use the energy provided by ATP to separate the two strands of the double helix (figure 5.2).[39]

The combination of these two activities enabled specialists to explain why the reverse gyrase is capable of introducing positive supercoiling in DNA.

The composite structure of the reverse gyrase also had an important meaning for the problem of the hot (or cold) origin of life. If the reverse gyrase is indeed indispensable to live in temperatures above 80°C, hyperthermophiles

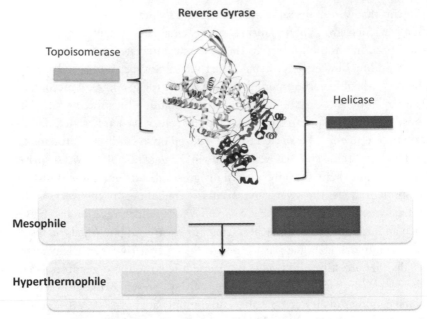

FIGURE 5.2. Origin of reverse gyrase. Upper figure: the reverse gyrase folded polypeptide chain with the DNA topoisomerase domains in gray and the helicase domain in black. Lower figure: emergence of reverse gyrase in hyperthermophiles by the fusion of the genes that separately encoded the topoisomerase and helicase domains in mesophiles.

could not have appeared until after the fusion between the DNA topoisomerase and helicase domains that had given birth to it. These two domains themselves are part of two large families of proteins, very ancient and very diverse. Under these conditions, hyperthermophiles could not be the direct descendants of the first forms of life on Earth, since they had to have been preceded by many organisms that had lived before a helicase and a DNA topoisomerase of those families fused to create the reverse gyrase. As we can imagine, this reasoning, which I published in 1995 in the specialized journal *Origin of Life and Evolution of the Biosphere*, didn't please the partisans of the hot origin of life.[40] It earned me a serious talking to by Stetter himself.[41]

Using the same reasoning, we see that it should be possible to ask this question: Was the last ancestor common to archaea and bacteria a hyperthermophile? We have seen that the response to this question was for me very important in choosing between different evolutionary scenarios. If the reverse gyrase is essential for life at very high temperatures, its presence in the last ancestor common to archaea and bacteria would indicate that this ancestor was indeed a hyperthermophile. By contrast, if it was absent in that ancestor, this would mean that the ancestor was a mesophile or a moderate thermophile.

How can we know whether the last ancestor common to archaea and bacteria possessed the reverse gyrase? I had undertaken to answer this question at the beginning of the 2000s by studying the evolution of this enzyme. At that time we had a dozen reverse gyrases of archaea and two sequences of bacterial reverse gyrase, those of *Thermotoga maritima* and *Aquifex aeolicus*. I then collaborated with Hervé Philippe to construct a phylogenetic tree from all these sequences. How was the tree of the reverse gyrase going to enable us to determine whether this enzyme was present in the common ancestor of bacteria and archaea? If it was present, we expected to observe two well-separated groups of reverse gyrase in the tree: one group corresponding to the reverse gyrases of archaea and another corresponding to reverse gyrases of bacteria. Each domain should have had its own version of reverse gyrase, to borrow Woese's terminology, since the latter would have evolved independently from each other from the time of the divergence between archaea and bacteria. But the results proved otherwise. In the tree built by Philippe, the two bacterial reverse gyrases were localized right in the middle of the reverse gyrases of archaea.[42] Thus there existed only one version of the reverse gyrase for all microbes from hell. These results, published in 2000 in *Trends in Genetics*, strongly suggested that the ancestor common to bacteria and archaea did not have the reverse gyrase: in other words, it was not a hyperthermophile. If that ancestor is connected to LUCA, as in Woese's tree, these results were in the direction of a mesophile LUCA, supporting the results published by Manolo Gouy the year before, based on the reconstitution of LUCA's 16S ribosomal RNA. For me, they had another significance. I had to definitely abandon my "simple" scenario of thermal reduction, leading from a cold LUCA to a hot ancestor common to bacteria and archaea, and adopt a more complex scenario with two independent events of thermoreduction (the one described in figure 4.7b, b2). As we saw in chapter 4, the complex scenario was also suggested by the independent adaptation of archaeal and bacterial lipids to function at high temperatures.

If reverse gyrase was absent in LUCA, it must have appeared either in bacteria or in archaea and was then transferred from one domain to another by gene transfer, which explains why all reverse gyrases resemble each other so much. Firm indications allow us to think that the reverse gyrase had first appeared in archaea before being transferred to bacteria. We had observed that the genes coding for the reverse gyrase in hyperthermophile bacteria *Thermotoga maritima* and *Aquifex aeolicus* are surrounded by other sequences that resemble genes coding for proteins of an archaeal type more than genes coding for proteins of a bacterial type. This is probably the signature of DNA fragments of archaea (blocks of genes) that were transferred to bacteria in very ancient times.

In 2002 the number of genes available was too small to be sure of this scenario. I resumed this work a few years later with Céline Brochier. Our new analyses, published in 2007 in the journal *Archaea*, confirmed our initial observations.[43] The branches leading to archaeal and bacterial reverse gyrases were still mixed in the phylogenetic tree constructed by comparing all twenty-eight sequences of reverse gyrases available at the time (figure 5.3a).

There indeed exists only one version of the reverse gyrase. But this work above all provided us with arguments to propose that the reverse gyrase was present in the last ancestor common to all archaea. When we constructed a phylogenetic tree including only sequences of archaea, we obtained a tree that was practically superimposable onto the phylogenetic tree of archaea that Brochier had obtained previously by adding end to end (concatenation) all the sequences of ribosomal proteins of archaea then available (figure 5.3b).[44] The fact that we could superimpose the tree of the reverse gyrase and that of ribosomal proteins suggests that the ancestor common to all archaea possessed this enzyme, since the sequence of the enzyme diverged from one archaea to another in the process of individualization of different species. On the other hand, the tree of bacterial reverse gyrases wasn't at all superimposable onto that of bacteria, obtained in this case by comparing their RNA 16S. Bacteria belonging to the same group were found on different places on the reverse gyrase tree. This confirmed our hypothesis according to which the reverse gyrase gene had been transferred, perhaps several times, from archaea toward several groups of thermophile bacteria.

Our analysis thus provided another strong argument in favor of a mesophilic LUCA, a last mesophilic or moderate thermophile bacterial common ancestor, and a hyperthermophile ancestral archaea, which corresponds to the result obtained a bit later by Mathieu Groussin, winner of the *La Recherche* Prize in 2012, by reconstructing the amino acids composition of universal proteins in LUCA.[45]

The appearance of the reverse gyrase and the invention of a new type of lipid would have been two major events of a particularly strong thermoreduction in the lineage leading to archaea. The ancestors of modern archaea would have thus been the first organisms on Earth to go beyond the 80 to 90°C barrier. It is an appealing idea. The ancestors of current archaea perhaps one day took refuge in the hottest of springs to escape competition with other organisms living in that distant time. That would have given the first archaea time to

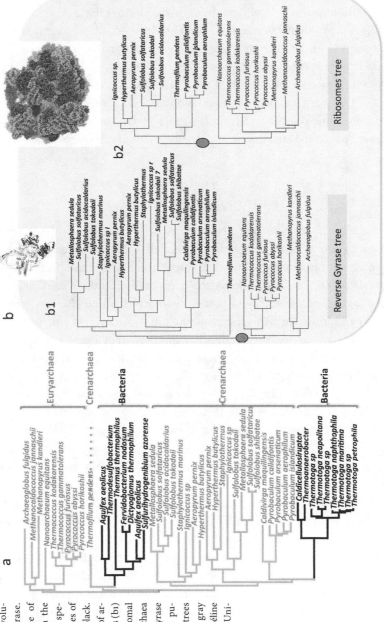

FIGURE 5.3. Evolution of reverse gyrase. A, phylogenetic tree of reverse gyrases with the names of archaeal species in gray and names of bacterial species in black. B, phylogenetic tree of archaeal reverse gyrases (b1) and of archaeal ribosomal proteins (b2; only archaea containing reverse gyrase are indicated). The putative roots of the trees are shown as dark gray circles. Courtesy of Céline Brochier-Armanet, Université de Lyon.

establish a safe home sheltered against their protoeukaryote predators or their bacterial competitors. We will return to this scenario at the end of this chapter.

Once the archaeal domain was well established, certain archaea would have stayed put, definitively attached to their original hell, whereas others would have managed to set off to explore cooler points of the planet, from the depths of the oceans to the soil of the forests and the villosities of our intestines. If this story has anything to do with reality, the reverse gyrase perhaps played the same role for archaea that some scientists believe the invention of language did in the appearance and development of our own species.

IS THE REVERSE GYRASE REALLY INDISPENSABLE IN HELL?

All the arguments I have just advanced are based on the hypothesis that the reverse gyrase is indispensable for life in hell—this hypothesis itself being based on the distribution of this enzyme in the living world. But is this hypothesis correct? In 2004 a Japanese scientist, Haruyuki Atomi, published in the *Journal of Bacteriology* a paper provocatively titled "Reverse gyrase is not a prerequisite for hyperthermophilic life."[46] Atomi had just perfected a technique enabling for the first time the elimination of a gene in a hyperthermophile. He had used as a model strain the naturally transformable *Thermococcus kodakaraensis*, which we discussed in chapter 2, and of course, as the first gene to eliminate, the one coding for the reverse gyrase, the only protein specific to microbes from hell. If the gene coding for the reverse gyrase is essential, the experiment should have ended in failure: it should have been impossible to isolate a mutant no longer possessing this gene that could still be capable of growing at 80°C. But Atomi had apparently succeeded, which justified the title of his paper.

A few weeks before this paper came out, I had learned this news from a Japanese colleague, Yoshizumi (Yoshi) Ishino, who had invited me to spend some time in Japan. He told me about it after we returned from a visit to the hot springs of Beppu on the island of Kyushu, where Zillig had isolated the first virus infecting a hyperthermophile: the SSV1 virus. After a day of tramping around this amazing site of dramatically colored hot springs in scenery worthy of Disneyland, we met up at nightfall in a burning pool, an *onsen*, one of the country's great attractions. It took me some time to be able to immerse myself gradually into this boiling pot; I admired Yoshi, who seemed perfectly comfortable in his bubbling bath. Once in the water, thanks to his encouragement, Yoshi turned to me with a serious look on his face. Now that we were on equal terms, naked in this giant cauldron, it was time to talk business.

"Do you know that a scientist from Imanaka's lab, Haruyuki Atomi, constructed a mutant of *Thermococcus* that no longer has the activity of the reverse gyrase and that this mutant still grows at 90°C?" Yoshi asked.

Those results were indeed counter to all my ideas on the essential role of this enzyme in microbes from hell. Was my analysis of comparative genomics going to be tossed away? Yoshi was a bit worried: How would I, already weakened by the unbreathably humid atmosphere of our burning bath, react to the news?

A few weeks later, while reading Atomi's paper, I was reassured. Between 60 and 85°C they had indeed observed only a weak slowing of growth in the mutant that no longer had the reverse gyrase. By contrast, at 90°C it grew two times slower than the wild strain, and at 93°C it didn't grow at all, whereas the nonmutated microbe did very well at that temperature. Contrary to what was asserted by the title of the paper published by our Japanese friends, the reverse gyrase is thus quite necessary to live at very high temperatures (at least above 90°C). These results confirmed the idea that this enzyme intervenes in one way or another to facilitate life in hell, since the destructive effect on the mutation in which the gene is deactivated intensifies as the temperature rises. To understand the provocative (and slightly erroneous) title of the paper by the Kyoto group, one must keep in mind that most Japanese scientists working with hyperthermophiles are fierce partisans of a hot origin of life.

All the same, since the mutant no longer possessing the reverse gyrase could still live (even poorly) at 90°C, why don't we find hyperthermophilic organisms without this enzyme in nature? Perhaps because the experiment carried out in the lab does not truly reflect the conditions that microbes from hell must confront in nature. It is possible that the absence of the reverse gyrase, if it doesn't prevent the growth of the isolated mutant in the laboratory, doesn't enable the latter to survive when it is in competition, in its natural environment, with microbes from hell possessing this enzyme. Furthermore, in the hot springs environment, in addition to high temperatures, we saw in chapter 3 that DNA is subjected to other stresses—for example, the presence of heavy metals, or radioactivity—that may be lethal at very high temperatures in the absence of the reverse gyrase (the experiment proving this assumption remains to be done). In any case, the mutant constructed by Atomi seemed very promising because its in-depth study should enable us to better understand the true role of the reverse gyrase, if we could understand why this mutant grew much less well than the wild strain between 80 and 90°C. Ten years later we still don't have the answer to that question, because nothing else was published by the Japanese team on that mutant. Furthermore, Qunxin She, a Chinese scientist working in Denmark, was unsuccessful in isolating a viable mutant of reverse gyrase in *Sulfolobus islandicus*.[47] In this microbe from hell, the reverse gyrase indeed seems essential. The role of the reverse gyrase, that truly atypical DNA topoisomerase, thus still remains one of the most frustrating of mysteries.

Archaea, the Origin of Sex, and the Size of Plants:
Another Story about DNA Topoisomerase

I have often given lectures with the title "Archaea: A Gold Mine for the Study of DNA Topoisomerase." This wasn't the opinion of one of the granters of funds I solicited in 1983 to study these enzymes in archaea. One of the arguments I had advanced to support my request was that "no one in the world has yet begun to study these enzymes." A serious mistake, according to an indiscreet colleague present at the meeting where the distribution of funds was being decided: one committee member had criticized my request by declaring, "If no one in the world has studied these enzymes, it's probably because they are not that interesting." In fact, at the time I myself didn't imagine just how interesting the DNA topoisomerase of archaea truly was. We've just seen one example of them with the reverse gyrase, discovered in 1984. Twelve years later, research on DNA topoisomerase of archaea would lead us by chance to a new discovery that this time was very dear to us, because it would teach us a lot about the origin of sex and about the size of plants. Here again, data from comparative genomics enabled us to make the connection between this archaeal enzyme and typically eukaryotic phenomena.

To understand how the study of the DNA topoisomerase of archaea led us to these unexpected discoveries in the molecular biology of eukaryotes, we must go back fifteen years. It is 1996 and I am seated at the computer in the library of the Institut de Génétique et Microbiologie at Orsay, consulting the database of the National Center for Biological Information (NCBI) in Bethesda, Maryland, which collects all the sequences of known proteins. Today, of course, I can do this from home, thanks to the Internet. In 1994, after discovering and purifying the type II DNA topoisomerase (Topo II) of the hyperthermophile archaea *Sulfolobus shibatae*, a scientist named Agnès Bergerat, who at the time was doing a postdoctoral fellowship at the lab, and a CNRS engineer, Danièle Gadelle, had just cloned and sequenced the genes coding for the two subunits (called A and B) of this protein (figure 5.4).

By translating the nucleotidic sequence of these genes through the genetic code, we obtained the amino acid sequences of these two subunits. We were curious to know if these sequences resembled more that of the Topo II of bacteria—the famous gyrase—or that of eukaryotes, which are both

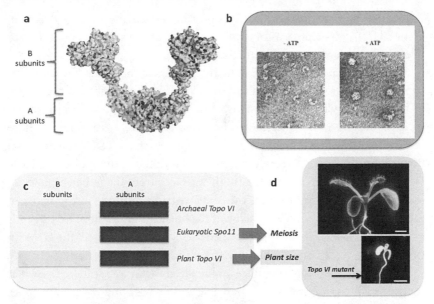

FIGURE 5.4. The fascinating evolutionary connections of archaeal DNA topoisomerase VI. *A*, atomic model of the open form DNA topoisomerase VI (Topo VI) from the archaeon *Sulfolobus* (courtesy of Dr. Danièle Gadelle, Université Paris-Saclay). The protein can capture two DNA double helices and force them to cross each other (see figure 3.4). The two B subunits can move, producing either an open or a closed conformation. *B*, electron microscope images of the closed and open conformations of *Sulfolobus* Topo VI, courtesy of Cyril Bulher. *C*, organization of the genes encoding the two subunits of archaeal and plant Topo VI and the eukaryotic protein SPO11. *D*, wild type (*top*) and Topo VI mutant (*bottom*) of the plant *Arabidopsis* at the same scale (courtesy of Dr. Sugimoto-Shirasu).

homologous. Agnès and Danielle had shown that the Topo II of *Sulfolobus* does not possess gyrase activity, just like the eukaryotic enzyme;[48] given the many points in common already demonstrated among archaea and eukaryotes, we expected the enzyme of the archaea to resemble more that of eukaryotes than that of bacteria. During this work we first obtained the sequence of the subunit A. When Agnès and Danielle began searching for the first time in the NCBI database for the proteins that this subunit most resembled, the results were surprising: the Topo II of archaea didn't resemble any other known Topo II, not that of bacteria, nor that of eukaryotes. It did resemble a yeast protein (a eukaryote) called SPO11, SPO being the abbreviation for sporulation. What was the relationship between a Topo II and sporulation? We were perplexed. Had we sequenced the right gene?

That day, seated in front of the computer, I launched another search into the NCBI database and again got SPO11, but the description of this protein (this is called annotation) had changed; SPO11 was now presented as a protein involved not only in sporulation but also in meiotic recombination. This was the eureka moment, since now it was possible to imagine a link between an archaeal DNA topoisomerase and SPO11, even if archaea have neither sex nor meiosis. To understand why, recall that meiosis is a very particular type of cellular division that results in the reduction by two of the number of chromosomes in sex cells (spermatozoa and ova in the human species). In the course of this process, whether in humans or in yeast, homologous chromosomes (one coming from the father, the other from the mother) recombine before separating.[49] The recombination of chromosomes (thus a DNA double helix) involves breaking the DNA into pieces, then assembling those pieces (a piece from the father, a piece from the mother, a piece from the father, etc.) to reconstitute entire chromosomes. This time we had a unifying thread between meiotic recombination, which involves the breakage of the two DNA strands of the chromosome, and a Topo II, since these enzymes must cut the DNA on the two strands to carry out its work (figure 3.4).

I dove into the recent literature and realized that it was still not known at that time which enzyme was responsible for the breaking of chromosomes at the onset of meiotic recombination. However, two research teams, including that of Alain Nicolas at the Institut Curie in Paris, had just demonstrated the presence of a protein fixed on DNA by a strong bond (covalent) at the extremity of the breaks that occurred during meiotic recombination.[50] For me there was no doubt: this protein had to have been SPO11. We already knew, thanks to genetics, that SPO11 intervened in meiosis, but didn't know at which stage. I called Alain Nicolas to give him the news:

"Hello, Alain. I know which protein breaks the chromosomes at the moment of meiosis."

"Are you back from a meeting in the US? Who made this discovery?"

"No, we're the only ones who know: Agnès, Danielle, you, and me, thanks to archaea . . . it's the protein SPO11."

It was, of course, just a hypothesis, but we were quickly able to test it by collaborating with Alain. A scientist on his team, Bernard De Massy, was able to show that the inactivation of the protein SPO11 in yeast through mutation indeed blocks meiosis—success!—and we published a paper in *Nature* in 1997.[51] Thanks, archaea!

The progress in genomics in the 2000s enabled us to find genes homologous to those coding for the two subunits of DNA topoisomerase type II of *Sulfolobus* in all archaea. We called this enzyme DNA topoisomerase VI (Topo VI), which related to the order of the discovery of different families of DNA topoisomerase. Topo VI corresponds, then, to the enzyme that resolves topological problems raised by the replication of the chromosome in archaea: in particular, separation of the two chromosomes at the end of replication.

The analysis of genomes also enabled us to find the gene SPO11 in all eukaryotes, animals, fungi, plants, and protists. The protein SPO11 is thus certainly responsible for the unleashing of meiosis in humans. And so there exists a strong evolutive link between archaea and the origin of sex of eukaryotes.

The analysis of the first eukaryotic genomes that started to be sequenced at the turn of the century had shown that there is no homologue of the archaeal Topo VI B subunit in animals and fungi (remember that SPO11 is the homologue of the A subunit). These eukaryotes thus do not possess a true Topo VI, but only SPO11 involved in meiotic recombination. By contrast, when the first genome of a plant, *Arabidopsis thaliana*, was published in 2002, it was noticed that the genome of this plant encodes homologues of the two Topo VI subunits.[52] We know today that all plants possess an archaeal-like Topo VI (figure 5.4). Just like other eukaryotes, plants possess the classic Topo II of eukaryotes. What was the purpose of the Topo VI of plants? The response was quickly provided by two European teams. Topo VI is essential in plants for a process called endoreduplication.[53] This process enables certain cells to have more than one pair of chromosomes (let's remember that we—*Homo sapiens*—possess twenty-three pairs of chromosomes). Instead of being simply duplicated, each chromosome is sometimes present in the cells of plants in four, eight, sixteen, or even thirty-two copies. In the absence of Topo VI, this process always stops at eight. But in plants, the number of chromosomal copies determines the size of the cells, which in turn determines the size of the plant itself. Mutants of plants no longer having Topo VI would thus have small cells and would therefore be small (figure 5.4). From this fact, plant mutants no longer having Topo VI are indeed viable, but in the form of bonsais.

Without archaea and their Topo VI, the Amazonian forest would thus resemble a completely ridiculous vast expanse of bushes. Without the Topo VI of archaea, all those magnificent plants that we admire every day wouldn't have the majesty that their great size confers on them. In 2014, it appeared that Topo VI also intervenes in the mechanism that enables plants to fix atmospheric nitrogen through the intermediary of bacteria that colonize the nodules present on the surface of roots.[54] In the absence of Topo VI, those nodules don't form. Topo VI is thus at the origin of plants that fix nitrogen, itself at the foundation of a large part of our agriculture.

Are Archaea Our Sisters or Our Mothers?

The amazing connections that can be made between an archaeal DNA topoisomerase, the origin of sex, and the size of plants has illustrated once again the unique evolutionary relationships between archaea and eukarya. Previously we saw that these two domains share very similar RNA polymerases and DNA replication proteins. I could have also mentioned the archaeal histone discovered in the early nineties by John Reeves of Columbus, Ohio, which is nearly identical to the well-known eukaryotic histones, small basic proteins that organize our chromosomes by winding DNA around them to form the so-called nucleosomes;[55] Steve Bell, a Scot now working in the United States; and the late Rolf Bernander of Sweden, who discovered proteins in archaea homologous to those that intervene in cell division and in the formation of membranal vesicles in eukaryotes.[56] All these proteins shared by archaea and eukaryotes either are not present in bacteria or are very divergent from their bacterial homologues. If we look again at Woese's tree, rooted between the bacteria and an ancestor common to archaea and eukaryotes, it is tempting to think that most of these proteins appeared in the branch that leads from LUCA to this ancestor (large grey arrow in figure 5.5a).

But some scientists want to go further (or, in my opinion, backward) to explain this proximity between archaea and us. After all, archaea are prokaryotes, so why not simply conclude that we (eukaryotes) descend from archaea, and thereby recover our good old view of evolution: from prokaryotes to eukaryotes. The proponents of this scenario, which I will henceforth call "the archaeal ancestor scenario," were encouraged by a paper published in *Proceedings of the National Academy of Sciences* with an evocative title: "The Archaebacterial Origin of Eukaryotes."[57] These authors had constructed a universal tree based on the addition end to end (concatenation) of sequences of amino acids of fifty-three proteins preserved in the three domains of life. In the trees they obtained, eukaryotes were no longer the brother group of archaea, but a subgroup of archaea, corresponding to one of the two great phyla recognized at the time: the Crenarchaeota (schematized in figure 5.5a).[58]

If we accept this scenario, we return *de facto* to the two-stage rocket from the mid-twentieth century. In the first stage, which originally corresponded to the diversification of bacteria (figure 1.6), the latter were simply replaced by archaea. Pushed out the door in 1977, the two-stage rocket returned through the window in 2008. It was not by chance that the authors of this study borrowed the old term "archaebacterial" for the title of their paper, bringing back the confusion between bacteria and prokaryotes. In this title, archaea reclaimed their bacterial status, the very one that Woese had attempted to eliminate once and for all. You can imagine that he was not very happy with these results. For me, it was a counterrevolution. Yes, of course, you may say, but revolutionaries might also be wrong. The twentieth century saw so many

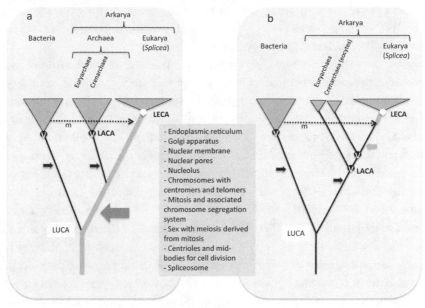

FIGURE 5.5. Are archaea our ancestors or sisters? *A*, the classical Woese tree in which archaea are monophyletic and sisters of eukarya. *B*, the "eocyte" tree in which archaea are our ancestors. In both trees, archaea and eukarya form a monophyletic group, the proposed "superdomain" arkarya (Forterre, 2015). The dotted lines correspond to the endosymbiosis at the origin of mitochondria. Bold lines in gray correspond to lineages in which all eukaryotic specific features (*middle panel*) might have appeared. Black arrows indicate the putative thermal reduction steps; black circles indicate thermophilic or hyperthermophilic ancestors. Note that in tree *B*, we share a hyperthermophilic ancestor with archaea.

revolutions gone astray that one can understand the attraction of counterrevolutions. It's true, but I am French, and many French people retain their love for our great Revolution, in spite of its excesses, and are cautious of counterrevolutions, as symbolized by Louis XVIII, a pot-bellied king suffering from gout who was brought to Paris in the irons of the enemy after Waterloo. I can see you frowning; a scientist should not be blinded by such ideological prejudices. You're right. I will have to argue on the scientific level.

First, I was not convinced by the results presented by Martin Embley and his team in their 2008 paper. When I looked at the individual fifty-three universal protein trees (obtained before their concatenation) in the supplementary material in the paper, I was puzzled by their topology. If eukaryotes indeed emerged from within archaea in all trees but one, it was never from the same archaeal branch. Sometimes they were a sister group of Crenarchaeota, sometimes of Euryarchaeota. In a certain number of cases they emerged from within Crenarchaeota or Euryarchaeota. Crenarchaeal or euryarchaeal species (or both) were also mixed in many trees. I concluded that these phylogenies were not significant. By adding up all these confusing phylogenies, the authors obtained an average tree in which eukaryotes fell more or less in the middle of archaea, a sister group of Crenarchaeota.

We saw in the previous chapter, when we were discussing the root of the tree, that the results obtained by molecular phylogenetics are sometimes very tricky when it is a matter of reconstructing very ancient events. In the problem discussed here, the situation is rendered difficult by the fact that the archaeal and eukaryotic proteins are both very similar to each other and very different from their bacterial homologues (remember the long branch leading to bacteria in the elongation factor trees of figure 4.5b). A great number of amino acid changes (mutations) must have occurred at the same position along the long bacterial branch. Today we observe a snapshot (the result of the last mutation); it is difficult to have confidence in the results obtained. Embley and his colleagues argued that they had used a new, particularly efficient method to retrieve a weak phylogenetic signal, but I was not convinced. In the case of one of the universal proteins—Kae1, which we studied in Orsay and which is involved in the formation of the universal modification of tRNA called t6A—Brochier obtained Woese's tree. In this tree, Euryarchaeota and Crenarchaeota were well separated. By contrast, with their method, Embley and his colleagues obtained a tree in which eukaryotes, which emerge in the middle of archaea, were a sister group of the hyperthermophile archaeon *Methanopyrus kandleri*, which is completely atypical.[59]

We will come back to the phylogenetic analyses of universal proteins at the end of this chapter, with data that confirmed my first impression. However, before that, I would like first to convince you that it is indeed extremely important to determine whether archaea are our ancestors. I will also discuss some of the biological arguments that make me very skeptical of the archaeal ancestor scenario and doubtful of the results obtained by some specialists in molecular phylogenies.

Why Is It Important to Determine Whether Archaea Are Our Ancestors or Not?

First, there is a question of nomenclature, because if archaea are our ancestors, that means that we ourselves are archaea, or that archaea don't exist (in any case, no more than fish). Let me explain. For modern evolutionists who followed the cladistics rules mentioned in chapter 4, which Hennig set up in 1966, a group of related organisms to which we give a name (a taxon) must be monophyletic—that is, all the descendants of the last common ancestor of this group must belong to it. Fish do not correspond to a monophyletic group because the common ancestor to all fish is also one of our ancestors, since vertebrates descended from fish. The term "fish" thus cannot be valid (and should not be used) for evolutionists, unless we consider that we are also fish. (I want to reassure you: you can continue to use this term at the fish counter of your grocery store.) It is amusing to note that the partisans of the scenario of the archaeal ancestor don't like the logical conclusion of this scenario— that is, that we are archaea. This is one of the reasons why they return to the term "archaebacteria." They are "gradists," a concept in evolution in which a living being always evolves from the simplest to more complex forms, with each stage in evolution corresponding to a "grade." In this view prokaryotes correspond to a lower grade (with eubacteria and archaebacteria), and eukaryotes to a higher grade. If we adopt this view, we no longer have to worry about the monophyly of the domains; archaea and eukaryotes form a monophyletic group. Yes, but they don't belong to the same grade. In fact, Woese himself flirted with gradism. For example, he never proposed giving a name to the clade grouping archaea and eukaryotes, whereas they indeed form a monophyletic group in his universal tree. This is probably because Woese really liked the trinity of life based on three domains that replaced the old dichotomy between prokaryotes and eukaryotes.[60] In a cladistics framework, this tree is also formally a dichotomy, this time between bacteria and the "superdomain" formed by archaea and eukarya. I recently submitted a paper for *Frontiers in Microbiology* in which, following a suggestion by Prangishvili, I propose christening this superdomain "Arkarya," a contraction of Archaea and Eukarya[61] (figure 5.5). Let's see this if this proposal will find any followers.

I sometimes enjoy proposing new names. Long ago, I suggested in a commentary to *Nature* replacing the terms "prokaryotes" and "eukaryotes" ("before the nucleus" and "true nucleus"), with their strong evolutionary connotations, with the neutral terms "akaryotes" and "synkaryotes," simply meaning "without nucleus" and "with nucleus."[62] More recently, I proposed replacing the name "Eukarya" with "Splicea."[63] For me, the name "Eukarya," derived from "eukaryote," is a bit too evocative of the term "prokaryote." Woese and his former student Norman Pace, the father of microbial molecular ecology, fought bitterly against the use of the term "prokaryote" because it mixes archaea and bacteria and leads to much confusion.[64] For example, viruses of archaea are often called phages, in reference to the bacteriophage name traditionally given to viruses of bacteria. I think we will never manage to rid ourselves of the term "prokaryote" if we don't rid ourselves at the same time of the term "eukaryote." I thus sought a new name for "Eukarya." For me, the most unique characteristic of this group is its "genes in pieces." We saw in chapters 1 and 2 that the mechanism enabling eukaryotes to manufacture messenger RNA in which information is coded continuously is called splicing (figure 1.3). Since their discovery, the origin of genes in pieces is the subject of debate among evolutionists, between the partisans of an ancient origin and those of a recent origin.[65] For the former, genes in pieces reflect the ancestral nature of genetic organization; for the latter, the current introns are the descendants of mobile elements that invaded eukaryotes—genes that, in the beginning, were not in pieces. This controversy helps us understand the second reason why it is so important to determine whether archaea are our ancestors or our "sister domain." If archaea are our ancestors, the debate about the origin of genes in pieces is definitely decided in favor of a recent origin, since the genes of archaea are not in pieces. If the archaeal ancestor scenario is correct, all the characteristics specific to eukaryotes listed in figure 5.5 necessarily appeared in a "prokaryote," in the lineage of archaea that gave birth to us (line in bold gray). But if archaea are our sister domain, as in Woese's tree, it is possible that some characteristics specific to eukaryotes are ancient (bold gray line in figure 5.5a) and were lost in the course of the evolution that led from the common ancestor of archaea and eukaryotes to the archaea (black arrows). For me, such losses could very well have happened, especially if the hypothesis of thermoreduction is correct, since "reduction" in biology is a synonym for "loss."

One might wonder if it is possible to lose a characteristic such as spliced genes with all the complex machinery (spliceosome) involved in splicing. In this case, a major discovery made ten years ago tells us that this is possible. This discovery was made by scientists studying atypical eukaryotic protists whose ancestors incorporated a red algae through endosymbiosis. In some of them, the endosymbiont conserved its nucleus in a reduced form, called a nucleomorph. Strikingly, the sequencing of several nucleomorphs' genomes has shown that, in some of them, the genes were no longer in pieces;[66] they had lost all introns initially present in the genes of the red algae. In addition, these genomes have lost all the genes coding for the proteins and the RNA that form the spliceosome. We can then imagine, for instance, that the RNA genes of LUCA were cut in pieces by introns, and that these introns were eliminated in the lineages leading to archaea and bacteria (possibly because of thermoreduction), while they were retained in eukaryotes. However, if archaea are our ancestors, this appealing scenario makes no sense.

Last but not least, the problem of the origin of eukaryotes is of primary concern to us not only as *Homo sapiens*, thus eukaryotes, but also as readers of this book. We saw in the previous chapter that the last archaeal common ancestor (LACA) had probably been a microbe from hell. Thus if we descend from archaea, LACA is also our ancestor (figure 5.5b). Are we, too, children of hell?

A Few Arguments against the Scenario of the Ancestral Archaea

Before going back to the phylogenetic analyses that have apparently supported the archaeal ancestor scenario, let's consider a few points that explain why I have always been very skeptical of this scenario. If archaea are our ancestors, we must first explain why their unique lipids have been replaced in eukaryotes by lipids of the bacterial type. We saw in chapter 3 that the lipids of archaea are very effective in all temperatures, and that they are less permeable to protons and sodium than those of bacteria and of eukaryotes, which is an advantage for conserving energy within cells.[67] Thus we can't see why these very effective lipids would have been replaced in eukaryotes by lipids that show no apparent advantage.

Another argument involves ATP synthases, about which I have spoken a lot in this book. We have seen that archaea possess a rotary ATP synthase on their cytoplasmic membrane, whereas in eukaryotes the related protein (the V-ATPase) is located in the membranes of intracellular compartments called vacuoles. Also, the eukaryotic V-ATPase functions are in reverse compared to the archaeal one, consuming ATP to force the passage of ions or protons through the membranes. In the archaeal ancestor scenario, we can't see what selective advantage might explain why the descendants of this archaeal ancestor would have transferred their ATP synthases from the cytoplasmic membrane toward the vacuoles and would have inversed its mode of functioning, thereby losing a formidable machine for manufacturing energy.

Another observation that, in my opinion, counters the archaeal ancestor scenario concerns viruses. There are many viruses infecting eukaryotes that have no viral homologue in archaea. In particular, whereas there are many RNA viruses infecting eukaryotes (AIDS virus, flu, chicken pox, Ebola virus, and so forth), we don't know of any RNA virus infecting an archaeon. If eukaryotes derive from archaea, this would mean that the RNA viruses of eukaryotes appeared recently, during the transformation of archaea into eukaryotes. This is puzzling, since RNA viruses seem very ancient, as discussed in chapter 4, possibly being the descendants of the first viruses that appeared at the time of the RNA world. Considering the instability of RNA at high temperatures, ancient RNA viruses were possibly eliminated when the ancestors of archaea adapted to higher and higher temperatures. When I proposed the thermoreduction hypothesis in 1995, I predicted that we wouldn't find an RNA virus in archaea. Twenty years later, this prediction has proved to be correct. Of course, such a prediction is always a bit dangerous. Everything could be challenged tomorrow if we discover an RNA virus infecting an archaeon.[68]

Ultimately, the archaeal ancestor scenario poses a fundamental problem: Is it possible to have one domain emerge out of another? All modern archaea have the same versions of universal proteins, whereas they have been present on our planet for probably more than three billion years. In all that time, they have evolved, they have diversified, but they are still archaea. The same might be said of bacteria. Referring to this observation, Woese wrote in a paper that appeared in 2000 in *Proceedings of the National Academy of Sciences*, "Modern cells are sufficiently complex, integrated and 'individualized' that further major change in their designs does not appear possible."[69] In other words, for Woese, as for me, an organism of one of the three domains will not be able to be transformed, even after a long period, into an organism of another domain. This is precisely what must have happened if archaea are our ancestors; we must imagine that a lineage of archaea distinguished itself from all the others to, at a given moment, free itself from the evolutive constraints that for three billion years prevented other archaea from becoming something other than archaea.

I know how the partisans of the archaeal ancestor scenario would have responded to the above arguments.[70] For them, it was the capture and domestication by an archaeon of a bacterium at the origin of mitochondria (dotted line in figure 5.5b) that completely transformed that archaeon, turning it into a eukaryote. They have proposed all sorts of hypotheses to explain how, from the association of an archaeon and a bacterium, all the eukaryotic characteristics listed in figure 5.5 would have gradually appeared. For instance, they proposed that the genes of the bacterium coding for the biosynthesis of lipids replaced those of the archaeon for the production of membrane lipids. To explain the emergence of genes in pieces, they suggested that spliceosomes originated from mobile elements called group II introns that were present in the genome of the bacterium at the origin of mitochondria. These mobile elements, which encode self-splicing RNA, are indeed evolutionarily related to the RNA molecules present in the spliceosome. In the proposed scenario, they invaded the genome of the archaeal host by cutting up its genes, leading to the formation of genes in pieces, and evolved progressively to form modern nucleosomes by recruiting proteins. To explain the emergence of eukaryotic viruses, they suggested that the viruses whose genome was integrated into that of the bacterium at the origin of mitochondria woke up to mix with those of the archaeon and participate in the formation of atypical viruses of eukaryotes, and so on. For me, all these ad hoc hypotheses are far-fetched. For example, some archaeal genomes indeed contain group II introns, but these mobile elements have never evolved in archaea to form even a protospliceosome. I must point out that spliceosomes (there are several slightly different versions) include several RNA molecules and a very large number of proteins (from several dozen to more than a hundred, depending on the species); some are bigger than the ribosome itself. How was mitochondrial endosymbiosis able to transform a simple group II intron made of a single piece of RNA into these monstrous spliceosome machines?

Some evolutionists give such importance to the endosymbiosis that gave birth to mitochondria in the origin of eukaryotes that they propose replacing the universal tree of life with a ring of life. In the ring scenario, the branch of archaea and that of bacteria fuse to form eukaryotes, a new version of the two-stage rocket. These models are still very popular, and it is assumed in many textbooks today that eukaryotes originated from the "fusion" of an archaeon and a bacterium.[71] Certainly, mitochondrial endosymbiosis played an important role in the formation of modern eukaryotes. Comparative genomics has shown that all current eukaryotes descend from a eukaryote that already possessed mitochondria.[72] Many genes coding for the bacterium at the origin of mitochondria were transferred from the chromosome of this bacterium to the chromosomes of eukaryotes. Most of the proteins coded by these genes still function in mitochondria, but today many others participate in the functioning of other cellular structures, which has introduced many changes in many aspects of eukaryotic cell physiology. But that must not have changed anything, or only in the margins, at the heart of the eukaryotic molecular biology systems that manage the functioning of the cell on the genetic level. However, drastic changes would have been necessary to transform an archaeon into a eukaryote, including the evolution of all archaeal versions of universal proteins into eukaryotic versions.

We don't know of any example of endosymbiosis that has led to such a reworking of the fundamentals of the molecular biology of the host, in particular of the structure of their ribosomes, and that has led to the creation of new versions of all its universal proteins. As proof I cite the endosymbiosis that gave birth to the chloroplasts present in plants. The capture of the cyanobacterium that is at the origin of chloroplasts led to the formation of a new branch of eukaryotes, the one that groups plants, green algae, and red algae, but all these eukaryotes have remained eukaryotes. This endosymbiosis has absolutely not modified the ribosomes of plants, nor has it created in plants new versions of all their universal proteins: they remain identical to those of all other eukaryotes!

More generally, the idea that mitochondrial endosymbiosis could have transformed an archaeon into a eukaryote seems to me completely improbable because it implies a sort of miracle that would have occurred only once. If the association of an archaeon and a bacterium was able to give birth to eukaryotes, why didn't this event occur several times in the three billion years that archaea and bacteria have cohabitated, giving birth to a number of new domains? Why don't we observe today bacteria living in archaea, or better, archaea/bacteria chimera that would have begun to be transformed into eukaryotes?

For all these reasons, it seems much more reasonable to me to think that the bacterium that gave birth to mitochondria was ingested not by an archaeon, but by a protoeukaryote (Woese's urkaryote) that probably already possessed many characteristics of modern eukaryotes. Comparative genomics has in fact shown that LECA, the last ancestor common to all modern eukaryotes, possessed not only mitochondria but also all the attributes of modern eukaryotes (cytoskeleton, genes in pieces, division by mitosis, sex with meiosis).[73] It is difficult to imagine that all these characteristics appeared in a very short amount of time: that is, between the moment when the archaeon/bacterium chimera was formed and the emergence of LECA (the small bold gray line in figure 5.5b). On the contrary, it seems plausible to imagine that most of them emerged in the branch that led from LUCA to modern eukaryotes by passing through protoeukaryotes (long bold gray continuous branch in figure 5.5a). Some possibly even appeared before LUCA, then were lost in archaea and bacteria, while others probably appeared only after the divergence of archaea and bacteria, in the branch specific to eukaryotes.

Let's look at a final argument that is often advanced by partisans of the two-stage rocket. In their opinion, the most ancient fossils found on the planet would be "prokaryote," 3.4 billion years old, whereas the first eukaryote fossils date from only 1.5 billion years ago for protists and 600 million years for multicelled organisms. It is very difficult to say if the microfossils, 3 billion or more years old, are archaea, bacteria, protoeukaryotes, or representatives of domains that no longer exist. Many biologists still have a tendency to confuse eukaryotes and "modern eukaryotes." Modern eukaryotes, the descendants of LUCA that all have mitochondria, are necessarily much more recent than bacteria, since evolutionists have been able to show that mitochondria descends from a particular subgroup of bacteria (we call them alphaproteobacteria), which had already diverged a long time ago from other bacteria at the moment of endosymbiosis. But the protoeukaryotes that preceded this endosymbiosis existed for a much longer time. Ultimately, even modern eukaryotes are perhaps much more ancient than we thought even a

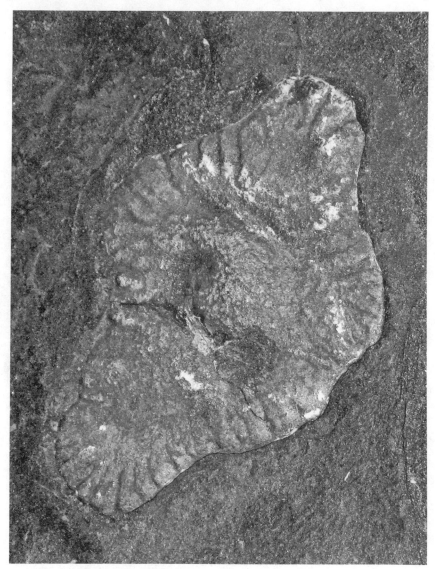

FIGURE 5.6. Macrofossils from Gabon. Picture of a macrofossil discovered in 2.1-Gyr-old sediments from Gabon. These mysterious multicellular organisms, visible to the naked eye, could be protoeukaryotes, protoanimals or witnesses of an extinct cellular lineage (El Albani et al., 2010).

few years ago. In 2010 a team from the University of Tours, France, discovered in the rocks of Gabon in Africa superb macrofossils (a few centimeters long) dating from 2.1 billion years ago, which indeed seemed to correspond to large multicellular eukaryotes, perhaps animals.[74] If that is the case, the eukaryote lineage could easily go back 3 billion years or more (figure 5.6).

Hunting the Superarchaeon

The evolutionists who support the idea that eukaryotes originated from the capture and domestication of the bacterium at the origin of mitochondria by an archaeon are aware of a major criticism put forward by their opponents: "We don't know of any example where an archaeon has domesticated a bacterium." To answer this criticism, they imagine an archaeon that is extinct today, a superarchaeon that perhaps had the ability to capture other cells through the movement of their cytoskeletons, a process called phagocytosis.[75] Significantly, the progress of metagenomic analysis during the last ten years has made it possible to search for such a superarchaeon among the plethora of new archaeal groups detected over the last two decades by molecular evolutionists using 16S ribosomal RNA as probes.

We will see at the end of this chapter that the superarchaeon, at the origin of eukaryotes, has not yet been found (I believe it does not exist). However, the analysis of DNA from new archaeal lineages detected in the environment by their 16S ribosomal RNA or the successful isolation of representatives of these lineages (or both) led to the discovery of new archaeal groups with more and more eukaryotic features, and even to a new major archaeal phylum. This is a fascinating story that continues to unfold.[76] I was involved in the beginning of that story through my collaboration with Céline Brochier and Simonetta Gribaldo. We found that some marine archaea discovered by DeLong in the 1990s, which were considered for a long time to be mesophilic Crenarchaeota, should in fact be considered members of a third major archaeal phylum. Let's see how this happens.

Originally, the marine bacteria discovered by DeLong were classified within Crenarchaeota because they were a sister group of these hyperthermophilic archaea in trees constructed from 16S ribosomal RNA. When, in 2008, Céline completed our archaeal tree based on ribosomal proteins (discussed earlier in the chapter) by adding *Cenarchaeum symbiosum*,[77] the first "mesophile Crenarchaeota" whose genome had been sequenced, we obtained unexpected results. Whereas *C. symbiosum* branched with Crenarchaeota in the 16S ribosomal RNA tree, it branched at the base of archaea in the ribosomal proteins tree, well separated from both Euryarchaeota and Crenarchaeota (figure 5.7a).

And so we proposed considering *C. symbiosum* (and other mesophilic archaea related to *C. symbiosum* by their phylotypes) as members of a third large phylum. We proposed naming these microbes Thaumarchaeota[78] (from the Greek *Thaumas*, "marvel" or "surprise") because we weren't expecting

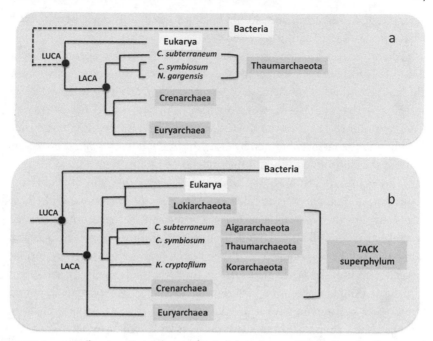

FIGURE 5.7. Different positions for new archaeal phyla in the tree of life. *A*, the archaeal tree rooted in the branch leading to Thaumarchaeota (adapted from Brochier-Armanet, 2008, 2011). The bacterial branch is represented by dotted lines because their sequences were not used in these analyses. *B*, eukaryotes branching with Lokiarchaeota (adapted from Spang et al., 2015) and sister group of the proposed TACK superphylum.

such an exciting result. We were able to publish it in the journal *Nature Microbiological Review*.

In the past few years, molecular ecologists have shown that Thaumarchaeota not only were abundant at the bottom of the oceans but populated all corners of the planet: the waters, the land, and even our skin. Thaumarchaeota thus have a lot of wind in their sails.[79] Most of them are able to oxidize ammoniac, and it seems that these archaea, still very little known, play a major role (perhaps the primary role) in the nitrogen cycle on our planet; until then we had thought that this role was uniquely held by bacteria. In 2011 a Japanese team discovered in the hot waters of a subterranean gold mine a new hyperthermophilic archaeon (encoding reverse gyrase) that they called *Candidatus* "Caldiarchaeum subterraneum." This archaeon branches at the base of the Thaumarchaeota in phylogenetic trees.[80] They suggested that this new archaeon could be the representative of a new phylum, Aigarchaeota.[81] I don't agree with these colleagues because *C. subterraneum* is always a sister group of other Thaumarchaeota in phylogenetic analyses (figure 5.7a).

I think that *C. subterraneum* should be considered as a hyperthermophilic Thaumarchaeota (a Thaum from hell).

Other microbes from hell that branch between mesophilic Thaumarchaeota and *C. subterraneum* have now been discovered in Yellowstone hot springs. This suggests that, as in the case of Crenarchaeota and Euryarchaeota, the ancestor of Thaumarchaeota was a microbe from hell. This conclusion again strongly supports the idea that LACA was itself a hyperthermophile, corresponding to the results obtained by Gouy and his students when they reconstructed the ribosomal RNA and the proteins of LACA.

Sequencing the genomes of *Cenarchaeum symbiosum* and other marine or terrestrial Thaumarchaeota, such as *Nitrosopumilus maritimus* and *Nitrososphaera gargensis*, revealed another major surprise. These genomes encode several eukaryotic-like proteins that are absent in other archaea, such as a homologue of the major eukaryotic type I DNA topoisomerase, an important target for antitumoral drugs in humans.[82] Their RNA polymerase large subunit is made of a single piece, as in bacteria and eukaryotes, whereas it is split into two smaller subunits in Crenarchaeota and Euryarchaeota. In several cases, thaumarchaeal genomes contain two proteins with similar functions that are otherwise present: one in Euryarchaeota, the other in Crenarchaeota.[83]

Why have Thaumarchaeota retained more eukaryotic-like features from LACA than other archaea? The archaeal tree with ribosomal proteins that Brochier obtained suggested an answer. When this tree was rooted using eukaryotic sequences, the root was located in the branch leading to Thaumarchaeota[84] (figure 5.7a). This suggests a parsimonious scenario in which all eukaryotic-like features specific to Thaumarchaeota were lost in a branch leading to a common ancestor of Crenarchaeota and Euryarchaeota.

However, it happens that some Crenarchaeota and Euryarchaeota also possess eukaryotic-like features that are missing in Thaumarchaeota.[85] For instance, some crenarchaeal genomes encode homologues of actin, one of the two major proteins in eukaryotes involved in the formation of the cytoskeleton. In that case, these eukaryotic-like proteins should have been lost in the branches leading to Thaumarchaeota but also to Euryarchaeota and to Crenarchaeota that do not contain this archaeal actin. With more and more archaeal genomes sequenced, it appears that many eukaryotic-like features are present in only a few archaea, suggesting that many gene losses occurred during the history of this domain. For instance, only a few thaumarchaea contain homologues of tubulin, the companion of actin in the formation of the cytoskeleton. These puzzling observations might be explained if LACA was more complex than modern Archaea; let's say a superarchaeon.[86]

This could be interpreted in two very different ways. In 2013 I wrote a paper for the journal *Archaea* with the title "The Common Ancestor of Archaea and Eukarya Was Not an Archaeon," in which I advanced all the biological arguments that I have briefly presented in this chapter.[87] For me, the complex LACA story tells us that modern archaea continued evolving through reduction, following the thermoreduction that occurred between LUCA and LACA. However, for proponents of the archaeal ancestor scenario, the presence of more eukaryotic-like characteristics in Thaumarchaeota, but also in some Crenarchaeota, in *Candidatus* "Caldiarchaeum subterraneum" (a thaumarchaeon for me, but an aigarchaeon for them) and in *Korarchaeum cryptofilum* (which Pace proposed as a member of a new phylum, a Korarchaeota) suggests that eukaryotes emerged within a superphylum grouping all these archaea. They call this superphylum TACK (Thaumarchaeota, Aigarchaeota, Crenarchaeota, Korarchaeota)[88] (figure 5.7b) and suggest that a superarchaeon of the TACK superphylum, capable of performing phagocytosis, once existed on our planet.

The grouping of eukaryotes with archaea of the TACK superphylum was apparently supported by phylogenetic analyses of universal proteins that appeared in several papers after 2010, building on the 2008 *Proceedings of the National Academy of Sciences* paper on the "archaebacterial" origin of eukaryotes.[89] However, other groups, using more or less the same universal proteins, obtained the opposite results, that is, Woese's classic tree, or concluded that it was not possible to decide between the two opposite topologies of the tree of life. This seems to lead us into a "phylogenetic impasse," an expression used by my colleague Simonetta Gribaldo at the Institut Pasteur.[90]

May 5, 2015: France Inter Studio—The "Tête au carré" Program

Yesterday I received an urgent phone call from a journalist at the France Inter radio station. I was supposed to go that afternoon to participate in a popular science program called "la Tête au carré" to talk about an article that was to appear the next day in the journal *Nature*. I knew about the article through the news that was circulating among the scientific community. Finally, the authors of this *Nature* paper might have put their finger on the assumed superarchaeon, a microbe that would be the sister of eukaryotes in the universal tree, breaking the phylogenetic impasse (figure 5.7b). This archaeon lived in sediments at the bottom of the Arctic Ocean, off the coast of Norway. The place where these sediments came from is near a grouping of hydrothermal hot springs called the Loki Castle, after the ancient Germanic god of fire, Loki. Apparently, this new archaeon was definitely a sister group of eukaryotes in phylogenetic analyses, putting an end to the debate. The microbe is really a superarchaeon since, according to Thijs Ettema, its large genome (around 5 million base pairs) encodes several new eukaryotic-like proteins never before seen in any archaeon. Furthermore, the functions of these proteins suggested that the Lokiarchaeon (as the authors called this new microbe) is able to build a complex network of intracellular membranes and has a cytoskeleton with an actin protein much more similar to the eukaryotic actin than those previously detected in other archaea. It is the ideal candidate for the engulfment of the bacterium that led to the mitochondria and triggered eukaryogenesis in the archaeal ancestor scenario. Sitting in front of a microphone at the radio station, I was not yet aware of all these details, and I was supposed to answer the journalist who was asking me to explain the importance of the forthcoming *Nature* article to a general audience. Was it true that the Lokiarchaeon (Loki for short) is the missing link between simple and complex life? Or, in the words of the title of the article that was to appear that day, do "complex archaea . . . bridge the gap between prokaryotes and eukaryotes"?[91]

I tried to explain the controversy surrounding our origins, mentioning that I remained skeptical of the archaeal ancestor scenario, but it was impossible to explain in ten minutes all my doubts, and I didn't have all the cards in my hand. My colleague Simonetta Gribaldo, who had joined me for this interview, was also a bit embarrassed. However, by revisiting the phylogeny of universal proteins, she had just obtained new results that seemed to support the archaeal ancestor scenario,[92] so it may be that Loki is the nail in the coffin of Woese's tree after all. In the coming days, Loki would create a buzz on the Internet and would soon have its own page on Wikipedia. If you Google Lokiarchaeota, you will be impressed by this scientific success story. The *Washington Post* headline was particularly striking: "Newly Discovered 'Missing Link' Shows How Humans Could Evolve from Single-Celled Organisms." Is that all?

At that point, the Loki story seemed to be a great achievement of the new metagenomic approach that I briefly mentioned in chapter 4. Indeed, the authors of the *Nature* article did not need to isolate and culture Loki to be able to sequence and analyze its genome. They had apparently been able to reconstruct this genome using powerful computer programs through multiple sequences of DNA fragments present in their sample. I tried to convey a message to the general public: this is a revolutionary approach that reveals a huge amount of microbial diversity that has up to now been hidden from us. All the same, it's important that the reconstruction of the genome be done correctly, and that's not so easy. The environmental samples that were analyzed always contained DNA from many different organisms; it is not at all clear that it could be sorted to reconstruct the genome under study without being contaminated by sequences originating in other organisms.

As you can imagine after reading this chapter, I was puzzled by the Loki story. After my radio interview, I was confused. Should I convert to the (Loki) archaeal ancestor scenario? I finally decided to take a look at the sequences of the Loki universal proteins myself. To distinguish between the different versions of universal proteins, it is useful to look at regions of the protein in which a few amino acids have been removed or inserted during evolution (as in figure 2.6a, white arrow). Specific insertions or removals (indels) have occurred in the lineages leading to each of the three domains; these can be used as evolutionary markers to trace the origin of an organism. If the Loki was really a missing link between archaea and eukaryotes, these sequences should have corresponded to a new version, in an intermediate position between the archaeal and the eukaryotic ones. I was surprised and confounded by the result of my investigations; most of the Loki sequences exhibited the typical archaea version of universal proteins, with no eukaryotic trace at all. In contrast, a few others exhibited the typical eukaryotic version (figure 5.8). I convinced two postdoctoral scientists working with me on the ERC EVOMOBIL project, Violette Da Cunha and Morgan Gaia, to join me in the analysis of the Loki sequences. We needed to be sure of the archaeal phylogeny if we wanted to understand the coevolution of archaea and their mobile elements, plasmids or viruses.

We started looking at all individual phylogenies of the universal proteins studied by Ettema and his colleagues, and their analyses confirmed my first impressions. In some trees, the Loki sequences branched within archaea, whereas in others, they branched in between archaea and eukaryotes (the missing link position), as in the tree obtained when you concatenate all universal protein sequences together. We concluded that the original Loki sample contained DNA traces from two different organisms: DNA from novel archaea belonging to a new lineage (the true Loki),[93] and DNA from unknown eukaryotes. These new eukaryotes possibly coexisted in the original Loki sample and their DNA was introduced by mistake in the Loki genome during the bioinformatics reconstruction, or maybe these eukaryotic genes were transferred from eukaryotes to ancestors of the Loki and were really present in the Loki genome. We will only know the answer when someone successfully cultures a Loki, enabling the sequencing of its true genome.

We were still puzzled by several questions, among them why these eukaryotic sequences branched in between archaea and "normal" eukaryotic sequences in phylogenetic analyses and not within eukaryotes? We found the

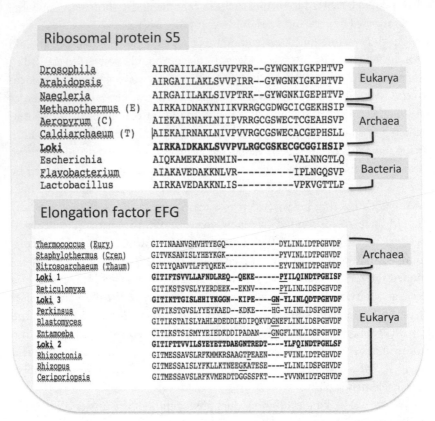

Ribosomal protein S5

Drosophila	AIRGAIILAKLSVVPVRR--GYWGNKIGKPHTVP	
Arabidopsis	AIRGAIILAKLSVVPIRR--GYWGNKIGKPHTVP	Eukarya
Naegleria	AIRGAIILAKLSIVPTRK--GYWGNKIGEPHTVP	
Methanothermus (E)	AIRKAIDNAKYNIIKVRRGCGDWGCICGEKHSIP	
Aeropyrum (C)	AIEKAIRNAKLNIIPVRRGCGSWECTCGEAHSVP	Archaea
Caldiarchaeum (T)	AIEKAIRNAKLNIVPVVRGCGSWECACGEPHSLL	
Loki	**AIRKAIDKAKLSVVPVLRGCGSKECGCGGIHSIP**	
Escherichia	AIQKAMEKARRNMIN----------VALNNGTLQ	
Flavobacterium	AIAKAVEDAKKNLVR----------IPLNGQSVP	Bacteria
Lactobacillus	AIRKAVEDAKKNLIS----------VPKVGTTLP	

Elongation factor EFG

Thermococcus (Eury)	GITINAANVSMVHTYEGQ------------DYLINLIDTPGHVDF	
Staphylothermus (Cren)	GITVKSANISLYHEYKGK------------PYVINLIDTPGHVDF	Archaea
Nitrosoarchaeum (Thaum)	GITIYQANVTLFFTQKEK------------EYVINMIDTPGHVDF	
Loki 1	**GITIFTSVVLLAFNDLREQ--QEKE------PYILQINDTPGHISF**	
Reticulomyxa	GITIKSTSVSLYYERDEEK--EKNV------PYLINLIDSPGHVDF	
Loki 3	**GITIKTTGISLHHIYKGGN--KIPE----GN-YLINLQDTPGHVDF**	
Perkinsus	GVTIKSTGVSLYYEYKAED--KDKE----HG-YLINLIDSPGHVDF	
Blastomyces	GITIKSTAISLYAHLRDEDDLKDIPQKVDGNEFLINLIDSPGHVDF	Eukarya
Entamoeba	CITIKSTSISMYYEIEDKDDIPADAN----GNGFLINLIDSPGHVDF	
Loki 2	**GITIFTTVVILSYEYETTDAEGNTREDT----YLFQINDTPGHLSF**	
Rhizoctonia	GITMESSAVSLRFKMMKRSAAGTPEAEN----FVINLIDTPGHVDF	
Rhizopus	GITMESSAISLYFKLLKTNEEGKATESE----YLINLIDSPGHVDF	
Ceriporiopsis	GITMESSAVSLRFKVMERDTDGGSSPKT----YVVNMIDTPGHVDF	

FIGURE 5.8. Amino acid alignment of two small regions of universal proteins showing significant indels. Sequences for the lokiarchaeal proteins are in bold. Top, indels in ribosomal proteins S5 that highlight the three versions of universal proteins (E: Euryarchaeota, C: Crenarchaeota, T: Thaumarchaeota). Bottom, indels in elongation factor G that highlight the eukaryotic affinity of the three different lokiarchaeal proteins.

answer when we noticed that the sequence of the Loki elongation factor EFG shared similar indels with a particular group of fungi (figure 5.8). When we added these fungi to the universal EFG tree, they branched in between Loki and eukaryotes instead of branching with other fungi within the eukaryotic part of the tree. This was clearly the result of the long branch attraction artifact (LBA) discussed in chapter 4. These fungi had especially long branches in our tree, indicating that they were rapidly evolving species. This is probably because they correspond to parasitic fungi. Indeed, the proteins of parasitic organisms often evolve much more rapidly than proteins of nonparasitic species, because of the continuous arms race between parasites and their victims.

We then realized that the same LBA phenomenon could explain the results obtained by the authors whose phylogenies supported the archaeal ancestor scenario. In their analyses, they always used the sequences of rapidly evolving archaea, such as the three microbes from hell discovered and isolated by Stetter: *Methanopyrus kandleri*, *Nanoarchaeum equitans*, or *Korarchaeum cryptofilum*. We had previously analyzed the phylogenetic positions of *M. kandleri* and *N. equitans* with Brochier and Gribaldo, and observed that they evolved more rapidly than other archaea.[94] In the case of *M. kandleri*, a methanogen that can grow up to 110°C, this can possibly be explained by the lack in this archaeon of a protein involved in the faithful transcription of the genome. In the case of *N. equitans*, this can be explained by the peculiarity of this fascinating organism. It is the smallest archaeal cell known, and it lives as a parasite straddling another microbe from hell, the crenarchaeon *Ignicoccus hospitalis*.

To determine the position of the true Loki in the universal tree of life, we decided to analyze the phylogeny of the largest universal protein (the one carrying potentially the most important phylogenetic signal): the RNA polymerase. We decided to use the same number of species in each domain (probably an important prerequisite to prevent unsampling bias) and to systematically avoid adding rapidly evolving organisms such as *M. kandleri*, *N. equitans*, or *K. cryptofilum* in our data set. The results were striking, as we obtained a very nice tree with the monophyly of all three domains and a very good resolution of all phyla inside each domain (figure 5.9a, b).

In this new tree, the true Loki turned out to be an early branching Euryarchaeota, whereas the archaeal tree turned out to be rooted in the branch leading to Thaumarchaeota, as in our previous analyses (figure 5.7a).

We now need to write our paper and analyze all possible universal proteins to confirm our initial results, but I am confident that we have found the right answer: Woese's tree is the correct one.

Before closing this chapter, now that we probably have the right tree in our hands (figure 5.9c), I would like to discuss two fundamental questions in evolutionary biology: one concerning the origin of archaea from hell, and the other concerning the origin of the three domains themselves.

In Hell . . . Safe from Raptors

If archaea are not our ancestors, they are certainly our close relatives. Why were their ancestors found in hell, whereas eukaryotes stayed in paradise? Perhaps to escape from us? Eukaryotes are great consumers of microbes. When we capture them, we see that not all have the opportunity to become endosymbionts; many end up as food that we consume in great quantities. Bacteria must have invented many mechanisms to defend themselves (the

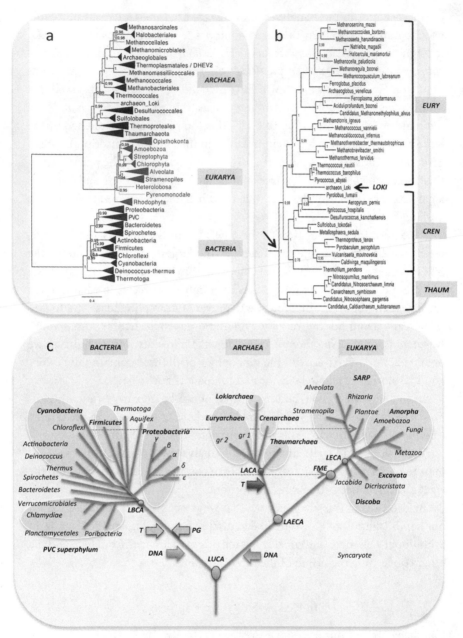

FIGURE 5.9. The universal tree of life. *A*, phylogenetic tree of the RNA polymerases showing the monophyly of the three domains. *B*, the archaea part of the RNA polymerase tree: the tree is rooted in the branch leading to Thaumarchaeota (as in figure 5.7a) and the Lokiarchaeon is an early branching Euryarchaeota (courtesy of Morgan Gaia, Institut Pasteur). *C*, a schematic view of the universal tree of life, adapted from Forterre, 2015.

production of toxins, for example), which probably explains why some bacteria much later became pathogens for humans—a passing of the baton, so to speak. Eukaryotic cells seem to have evolved by becoming bigger and bigger, the easier to eat prokaryotes. Chuck Kurland, one of the pioneers of molecular biology and the study of ribosomes who works in Sweden, compared eukaryotes to raptors, those terrifying dinosaurs from the film *Jurassic Park*. To escape the raptors, bacteria and archaea would have evolved to increasingly small sizes and increasingly rapid cell divisions.[95] This raptor theory can be combined with that of thermal reduction: it was perhaps also to escape from protoeukaryote raptors that the ancestors of archaea and bacteria took refuge where the raptors could not survive: that is, in hell, where the temperature was constantly above 60°C.

When the Tempo of Evolution Changed

Let's now look at the final question: How were the three different versions of universal macromolecules, proteins, and ribosomal RNA created? From my point of view, the best hypothesis was again proposed by Woese and Fox in several papers.[96] In their opinion, the speed of the evolution of macromolecules was faster at the time of LUCA than it became after the establishment of the three domains. This accelerated evolution led to the rapid formation (in a few tens or hundreds of millions of years) of different versions of universal macromolecules during the divergence of lineages that led to archaea, bacteria, and urkaryotes. Then the speed of evolution of macromolecules slowed, which fixed the present versions in the respective ancestors of archaea, bacteria, and eukaryotes. And so these versions remained practically unchanged for close to three billion years.

Several hypotheses have been proposed to explain this slowing in the tempo of evolution. In 2006 I suggested that the transition of RNA genomes toward DNA genomes, possibly triggered by viruses, could have contributed to this slowing.[97] In fact, the mechanisms of replication and reparation of DNA are more sophisticated than those of RNA, which enables a more faithful replication of DNA genes. I thought for a time that this transition could have taken place three times, once at the origin of each domain. I now think that it took place only twice: once in the branch of bacteria and once in the branch leading to Archaea and Eukarya (Arkarya, so to speak). This would explain why the version of universal proteins in bacteria is so divergent from that of Arkarya, whereas the archaeal and eukaryal versions are so similar. Indeed, if the common ancestor of Arkarya had a DNA genome, the tempo of evolution was already much slower at the time of the divergence between the lineages leading to Archaea and Eukarya.

September 4, 2015: Santiago, Chile—Thermophile 2015 Meeting

In my hotel room, I am finalizing the pictures for this book. I should send them to the editor before leaving for a trip in the desert north of Chile to visit the geysers and hot springs of El Tatio, 4,200 meters above sea level. Study-ing microbes of hell has many advantages. Regular meetings on this topic are often organized in nice countries with a tradition of working on these bugs. There are also many hot springs in Antarctica that are studied by Chilean microbiologists led by Jenny Blamey, who has organized this meeting. Today is a celebration for my wife, Évelyne, who shared with us her deep-sea experi-ence in chapter 2. She got one of the four best poster prizes of the meeting. Such prizes are usually reserved for young students and postdoctoral fellows. As a senior scientist, Évelyne is proud to share the honor with three of them! Her poster has caused a buzz at the meeting; she has presented some excit-ing electron micrograph pictures showing novel types of interaction between microbes of hell that she had taken a few days before in Paris. One of them shows two hyperthermophilic bacteria (two *Thermotoga*) that have apparently captured inside their toga one of the *Thermococcus* (an archaeon) that Évelyne brought back from her 1999 AMISTAD expedition. One can see a strange appendage that seems to connect the two microbes. Stetter was thrilled by this observation and was calling all his friends to have a look at the poster. Hey, what's going on there? Two bacteria eating an archaeon? Fight—or dinner—in hell? Microbes from hell are still full of surprises. You just have to put two of them together in the same room and something happens.

Yesterday I gave a presentation at the Santiago meeting in which I re-viewed Prangishvili's work on archaeal viruses: their origin and evolution. I also presented Violette and Morgan's last analyses on the position of the Lokiarchaeon and the universal tree of life. This is still a controversial topic, but I am now confident that we have the right answer. In my opinion, there is nothing more important in biology today than determining the topology of the universal tree of life—that is, understanding the history of living beings on our planet (a very personal opinion, of course). In the course of the meet-ing, I talked to Hiroki Higashibata, a colleague who spent a year in our Orsay laboratory trying to isolate reverse gyrase mutants. He is back in Japan now, following up on the work he began in France. Maybe we will finally under-stand from his work the biological role of reverse gyrase, the only protein specific to hyperthermophiles. I also used this opportunity to talk to another

Japanese colleague, Yoshizumi Ishino, to help him prepare for his visit to Paris for a sabbatical at the Institut Pasteur (he is the one who introduced me to the hot springs in Beppu, Japan). Yoshi will try to reproduce in a test tube the mechanism that initiates chromosome replication in archaea.

I love meetings where you have the chance to connect with colleagues and friends from all over the world. This time, I have a present for them: I brought copies of the 1897 paper of Bradley Moore Davis, the botanist who wrote the poem found in the epigraph to this book. I discovered Davis's paper while preparing the references for this book. I realized that no one in the present scientific community knew him and his pioneering work. This is not surprising, since very few scientists know the history of their branch of science. It's amazing to discover that this nineteenth-century gentleman anticipated the work of Zillig and Stetter (and many others after them) when they took up the challenge of isolating and studying microbes living in Yellowstone hot springs. By the way, I somewhat regret that I must now stop working on this book. Maybe, following in Davis's footsteps, I should have called hyperthermophiles "children of steam and scalded rock" instead of "microbes from hell"; it's definitely more poetic. But when I look at a picture of Évelyne, apparently with a poor *Thermococcus* being held prisoner between two *Thermotoga*, it seems they do have a diabolical side after all!

Epilogue

Our journey into the world of microbes from hell has led us to explore many questions, many of which still have no answer. It is part of science's charm never to give ready-made answers. We have nonetheless acquired some facts that guide us in our quest for the past. For example, we know that one day, a very long time ago, probably more than three billion years ago, an event of recombination (a mixture in some way) of two genes present on the chromosome of a thermophile ancestor of archaea produced a new protein, the reverse gyrase. This single event, in an isolated individual, had considerable repercussions for the history of life on our planet. The invention of the reverse gyrase was perhaps at the origin (at least in part) of one of the three domains of current living beings. Similarly, three billion years later, the appearance in *Homo* of mutants capable of better articulating sounds enabled our species, through language, to explore new spaces of communication, to create cultures, and, for some, to look at the history of life itself, finally discovering the reverse gyrase.

Each time I've used the word "invention," it was of course a manner of speaking. The new processes that appear in the course of evolution are the fruit of mutations that are produced by chance and are either eliminated, if they are harmful, or maintained through natural selection, if they provide an advantage to the individuals that carry them. For example, the association of two proteins that led to the formation of the reverse gyrase was accidental. It occurred following one of these multiple recombinations affecting a DNA molecule that marked the entire history of life (and is still marking it). That day, the association was fruitful. The microbe that inherited that genetic "accident" acquired a "plus" that enabled it to function a bit better than its neighbors and to transmit the reverse gyrase to its descendants. These were able to explore new shores and live a new life sheltered from any competition with other microbes that couldn't follow them into hell.

The appearance of microbes from hell was one of the great moments in the history of life on our planet. While studying them, we have demonstrated phenomena of adaptation that are exceptional on a molecular level: the double membranes that fuse and superhelices of DNA that turn in the opposite direction. Thus we understand better the universe that surrounds us. All living beings on Earth are connected by the same evolutionary thread. The same great mechanisms are at work; they have enabled—and still enable—life to adapt to an extraordinary variety of living conditions. The microbes that live in hell are made up of the same molecules as we are. We are all stardust. By studying hyperthermophiles we set off in search of our roots; we have most likely met our cousins, now quite distant, but who are without a doubt members of our family.

Some might ask, "Is it that important to retrace the most ancient history of life on our planet?" By briefly retracing the debates that shake up the scientific community, in particular concerning the origin of eukaryotes, we have been confronted with many hypotheses and much speculation. We don't really have a choice about that when we try to clarify the mystery of our origins. If we study a mechanism at work today in the biosphere—for example, the role of the reverse gyrase—we can always hope to arrive, following decisive experiments that leave no place for doubt, at consensual conclusions. This isn't the case when we study evolution. Without a time machine, it is impossible to reconstitute all of our past, because many archives have disappeared. We have to call on our imaginations, and we know that they can play tricks on us. Some scientists refuse to play this little game, believing that there are indeed enough pure, hard phenomena to study without wasting time dreaming. I think that's a mistake. Scientists also need to dream to live. More prosaically, they need to dream to work (work itself not always being a pleasure: experiments are not always carried out at the foot of volcanoes, even when you're studying microbes from hell). And dreams sometimes lead to major discoveries.

One of the greatest scientists of the last millennium, Guillaume Gilbert, who died in 1603, spent his life rubbing together every material he came across to determine whether the rubbed matter could attract a needle balancing on a pivot. This great dreamer wanted to verify and understand the ancient Greeks' observation that amber (*elektra* in Greek), when rubbed together, attracts particles. I rub this, I rub that, I attract this, I attract that, and so on, and I write down in my notebook the results of my observations. For fifty years Gilbert, who was also the official physician of Queen Elizabeth I of England, unwittingly spent his spare time founding the science of electricity.

That work, which he described in a wonderful book, *De Arte Magnetica*, was taken up and developed by a whole lineage of illustrious physicists, most of whom are forgotten today (with a few exceptions, such as Benjamin Franklin in the eighteenth century and, closer to us in time, Albert Einstein). How many of his intelligent contemporaries must have asked our dear Guillaume, "What's the use" of rubbing things together night and day? From time to time, politicians still ask this sort of question about basic research (which is still called academic with a bit of condescension). Gilbert could not have answered that he was opening the door to high-speed trains, DVDs, the washing machine, the television, the cell phone, and even the atomic bomb.

It must have been a bit boring to go around rubbing things all the time, in between visits to the queen. He must have been truly motivated. And he was. Gilbert thought that the force that attracted small particles in suspension toward the rubbed amber was of the same nature as that which attracted Earth and the moon. This was a major error, because he was confusing electromagnetic force with gravity, but the important thing is that he had an idea that pushed him to work during the days when the queen didn't need his services. I sometimes tell my students that what's most important is not having correct ideas, but just having ideas: "Work for ideas. Even if they're wrong, something will always come out of them." To think that we have understood one of the secrets of nature is what encourages us to work and supports us when the experiment has failed for the umpteenth time. For example, since I am now studying viruses, it's just as well to think they played a starring role in the great saga of evolution (and I sincerely think this). With that frame of mind, science is a true pleasure.

We're living in troubled times. People use the tools that science has put at their disposal to organize massacres and blow each other up in the name of the irrational. We all have a great responsibility—scientists, journalists, politicians, educated men and women—not to leave the next generations to flounder and not know of the marvels of nature. Among these marvels some are quite visible to all, from the polar bear to the statues of ancient gods, while others are more hidden, present in unexpected places. What a pleasure and what a feeling to discover them, to decode their secrets and understand their message after the billions of years that have seen life emerge and unfold on our planet.

Acknowledgments

This book is dedicated to Carl Woese, who changed my life and those of many other scientists by opening up a completely new world in biology: the concept of three—not two—biological domains. It is also dedicated to Wolfram Zillig, who introduced me to microbes from hell. And I salute Karl Stetter, who gave microbes from hell their name—hyperthermophiles—and provided us with new, fascinating bugs nearly every year for close to twenty years. Thanks to Daniel Prieur, who organized the AMISTAD expedition and invited Évelyne Marguet to participate, and to Christian Jeanthon, who supervised the expedition.

I am especially grateful to Stephane Frey, who encouraged me to write the first version of this book (in French) and supervised its publication. This English version was initiated by Christopher Chung; thanks to him and to all the University of Chicago Press team, Evan White, and Christie Henry for this great opportunity to update this book and publish it in the international language of our day. It was a pleasure to work with Teresa Lavender Fagan on the translation.

I am grateful to the many colleagues who provided me with pictures for the illustrations, many from their personal collections. Thanks to Violette Da Cunha, Sukvinder Gill, Aurore Gorlas, and Évelyne Marguet for reading and correcting the updated French version. Thanks to all past and present members of the BMGE teams at Orsay and the Institut Pasteur, some of whom joined us after reading the first (French) version of this book, and who have performed great work with our favorite microbes.

A special note of gratitude to David Prangishvili, the lord of archaeal viruses, who agreed to come with me to the Institut Pasteur to continue digging into the hidden world of those viruses from hell. These scientific adventures were made possible thanks to André Berkaloff and Philippe Kourilsky, who invited me to set up research groups when they were directors of the Institut de Génétique et Microbiologie in Orsay and the Institut Pasteur in Paris, respectively. The work of the BMGE teams has been supported by many grant agencies. A special thank-you to the European Research Council (ERC) which supports our present work.

Finally, thanks to my mother, who read and corrected the first version, and to my wife, Évelyne Marguet, who also spent her spare time on the book and provided me with the exciting account of her dive in the *Nautile*.

Notes

Chapter One

1. For the history of phage therapy, see the fascinating book *Viruses vs. Superbugs* by German journalist Thomas Hausler (Hausler, 2007).

2. In 1990, the Australian microbiologist John Fuerst discovered that a poorly studied group of bacteria, Planctomycetes, contains intracellular membranes, some of which form a pseudonucleus around their chromosome. Similar structures have now been found in other bacteria related to Planctomycetes, such as Verrucomicrobiales (Devos et al., 2013).

3. Davis, 1897; Setchell, 1903.

4. Brock and Freeze, 1969; for a review of the life and achievements of Thomas Brock, see Brock, 1995.

5. Brock et al., 1972.

6. Zablen et al., 1975.

7. Woese and Fox, 1977a; for a historical account of Woese's discovery, see science historian Jan Sapp's book *The New Foundations of Evolution: On the Tree of Life* (Sapp, 2009) and journalist Tim Friend's book *The Third Domain: The Untold Story of Archaea and the Future of Biotechnology* (2007).

8. The term "LUCA" was invented during a meeting on the last common ancestor, organized in France in 1996 at the Fondation des Treilles. For a historical account, see http://www-archbac .u-psud.fr/Meetings/LesTreilles/LesTreilles_e.html/.

9. Woese et al., 1980.

10. Magrum et al., 1978.

11. Darland et al., 1970.

12. For an early review of Woese's work, see Fox et al., 1980.

13. Kandler and König, 1978.

14. Zillig et al., 1978, 1979; Prangishvili et al., 1982.

15. Huet et al., 1983.

16. Forterre et al., 1984.

17. Woese et al., 1990.

18. For a historical account of the discussions between Woese and his colleagues leading to the term "domain," see Sapp, 2009.

19. Cacace et al., 1976; Zillig et al., 1979. The team from Naples missed the small subunits of the archaeal RNA polymerase in their purification and thus confuse its two large subunits with those of the bacterial enzyme.

20. Stetter et al., 1981.

21. Zillig et al., 1981.

22. Zillig et al., 1982.

23. Zillig et al., 1983a.

24. Prangishvili et al., 1982.

25. Zillig et al., 1983b.

26. These two branches correspond to the phyla known today as Euryarchaeota and Crenarchaeota; see chapter 4.

27. Basen et al., 2014; for the discovery of *Pyrococcus furiosus* see Fiala and Stetter, 1986.

28. Stetter, 1982; Stetter et al., 1983.

29. Stetter, 1989; for a historical account by the main protagonist, see Stetter, 2013.

30. Baross and Deming, 2003.

31. Corliss et al., 1981.

32. Trent et al., 1984.

33. White, 1984.

34. Stetter et al., 1993.

35. Kurr et al., 1991.

36. Blöchl et al., 1997.

37. Kashefi and Lovley, 2003; for a public discussion of these results, see criticism by Cowan, 2004, and response by Kashefi, 2004.

38. Takai et al., 2008.

39. Huber et al., 1986.

40. Huber et al., 1992.

41. Crick, 1962. Three successive nucleotides in an RNA or DNA message specify a specific amino-acid; see chapter 3.

42. Janekovic et al., 1983.

43. The virus SSV1 (originally named SAV1) was first detected as a free UV-inducible plasmid in *Sulfolobus* cells and later as lemon-shaped particles (Martin et al., 1984). It was renamed SSV1 because its host turns out not to be a *Sulfolobus acidocaldarius* strain, but a new species, *Sulfolobus shibatae*. For the first demonstration of SSV1 infectivity, see Schleper et al., 1992.

44. Prangishvili et al., 1999.

45. Arnold et al., 2000. Rudiviruses and lipothrixviruses have now been grouped together in the same order, Ligamenvirales, based on the structure of their major virion proteins (Prangishvili and Krupovic, 2012).

46. Oparin published his first book on the origin of life in 1924 in Russian. The book was translated into English under the title *The Origin of Life on the Earth* (New York: Academic Press, 1957). See also Oparin, 1959.

47. Miller, 1953.

48. Corliss et al., 1979.

49. Gold, 1992.

50. Weigel and Adams, 1998.

Chapter Two

1. Woese, 1981.

2. The concept of operon was proposed in 1961 by the French Nobel laureates François Jacob and Jacques Monod (Jacob and Monod, 1961).

3. This "intron early" hypothesis has been opposed by proponents of an "intron late hypothesis" in which introns originated from bacterial mobile elements brought to eukaryotic cells by the bacterium at the origin of mitochondria (discussed in chapter 5). For original papers discussing the puzzling organization of eukaryotic genes and intron early hypothesis, see Gilbert, 1978; Gilbert et al., 1986. For a recent review of the subject with alternative viewpoints, see Irimia and Roy, 2014; Rogozin et al., 2012; Rodríguez-Trelles et al., 2006.

4. At that time (1978) DNA gyrase had been recently discovered by the great American molecular biologists Howard Nash and Martin Gellert (Gellert et al., 1976).

5. Drlica, 1992; for a recent review on the role of supercoiling in bacterial transcription, see Dorman, 2013.

6. Oparin, 1959; translation of his book published in Moscow in 1957 (first edition in 1924). I read the 1965 French translation by P. Gacaudan and M. Guyot, *L'origine et l'évolution de la vie*.

7. Forterre et al., 1984.

8. For a comprehensive textbook by the person who discovered topoisomerases, James Wang, see *Untangling the Double Helix* (Wang, 2009). For a recent review more oriented toward the diversity and evolution of DNA topoisomerases, see Forterre, 2011a.

9. The enzyme was first named and described by our Japanese competitors (Kikuchi and Asai, 1984) as a multimeric type II DNA topoisomerase because it was contaminated by the RNA polymerase. We could show that reverse gyrase was indeed a monomeric type I DNA topoisomerase (see chapter 3 for the difference between type I and type II; Forterre et al., 1985). Our Japanese colleagues confirmed this result (Nakasu and Kikuchi, 1985).

10. Despite being among the most thermotolerant animals, Alvinella are nevertheless killed by exposure at temperatures between 50 and 55°C (Ravaux et al., 2013).

11. From samples taken during the AMISTAD expedition, Marteinsson isolated two thermophilic bacteria, *Marinitoga thermophila* and *Rhodothermus profondi*. The first is a deep-sea relative of the terrestrial *Thermotoga* discovered by Stetter on the island of Vulcano (Alain et al., 2002; Marteinsson et al., 2010). Growth of this bug is enhanced by hydrostatic pressure, a characteristic of piezophilic organisms. The second one is a relative of *Thermus*, discovered earlier at Yellowstone.

12. Huber et al., 1990.

13. The study of Christian's traps revealed that the archaeal community composition was distinct from vent to vent within the same vent field and varied within short periods of time. It also showed that the relative abundance of archaeal populations represented from 14 to 33% of the total microbial community recovered in the collectors (Nercessian et al., 2003).

14. Edmond Jolivet later isolated the most radioresistant of known hyperthermophiles, *Thermococcus gammatolerans*, to be discussed in chapter 3 (Jolivet et al., 2003).

15. This is no longer true; we will see by the end of this chapter that Prangishvili's student Ariane Bize has shown, for instance, that rudiviruses, once believed to be gentle viruses, in fact kill the cells. Other viruses from hell, such as ATV, also turned out to be efficient cell killers.

16. Lepage et al., 2004.

17. I am speaking here of *Thermococcus nautili*, a strain containing several plasmids and producing viral vesicles (Soler et al., 2010; Gaudin et al., 2014). This strain would finally be described and sequenced in 2014 (Oberto et al., 2014, and reference therein).

18. Geslin et al., 2003.

19. Gorlas et al., 2012.

20. Krupovic et al., 2014.

21. Beveridge, 1999; for a more recent review on bacterial membrane vesicles, see Kulp and Kuehn, 2010.

22. Soler et al., 2008; for a review and nanotube pictures, see Marguet et al., 2013.

23. Soler et al., 2010.

24. Deatherage and Cookson, 2012.

25. Izquierdo-Useros et al., 2011.

26. Melo et al., 2015; see also Théry, 2015.

27. Valadij et al., 2007.

28. Winans, 2005.

29. Shetty et al., 2011.

30. Dubey and Ben-Yehuda, 2011.

31. Marguet et al., 2013.

32. http://www.isevmeeting.org/.

33. Nelson et al., 1999.

34. Soler et al., 2008.

35. Gaudin et al., 2013.

36. Gaudin et al., 2014.

37. Soler et al., 2015.

38. Meckes, 2015.

39. La Scola et al., 2008.

40. Zillig died in 2005 after a long fight against prostate cancer, one month before an international meeting entitled "Archaea: The First Generation" was held in Munich to honor two pioneers of archaeal research: Wolfram Zillig and Karl Stetter. For a review of Zillig's legacy, see Albers et al., 2013.

41. For a comprehensive review of Prangishvili's career, "The Wonderful World of Archaeal Viruses," see Prangishvili, 2013.

42. Häring et al., 2005b; Bettstetter et al., 2003.

43. Häring et al., 2005a.

44. Bize et al., 2009.

45. Quax et al., 2011.

46. Forterre and Prangishvili, 2009, 2013.

47. Dupressoir et al., 2012.

48. Cortez et al., 2009.

49. Raoult et al., 2004.

50. I recently realized that it's very difficult, if not impossible, to define "what is a living organism?" Think about the history of mitochondria, cellular organelles that were once intracellular bacteria. All biologists agree that bacteria are living but mitochondria are not (they are organelles). However, it is impossible to precisely define when the transition between "living" and "nonliving" took place during the evolution leading from an intracellular bacterium to a mitochondria.

51. To know more about the virocell concept, see Forterre, 2013a, and references therein.

Chapter Three

1. In reality, the image of electrons rotating around the nucleus of atoms like planets around the sun is misleading. They form a "cloud" in which the precise position of any electron at any moment cannot be determined.

2. White, 1984.

3. Rice et al., 1996.

4. For a review of hyperthermophilic proteins, see Unsworth et al., 2007.

5. Saiki et al., 1988.

6. For a review of extremenzymes, see Elleuche et al., 2014.

7. As an example, the structure of the DNA repair protein XPD, involved in cancer suscepibility in humans, was first obtained for a *Sulfolobus* protein (Liu et al., 2008). This helps to explain why some mutations of this protein can favor the development of cancer.

8. Watson and Crick, 1953a; for a fascinating historical account of DNA studies, see *The Path to the Double Helix* by the historian of science Robert Olby (1974; revised 1994).

9. As a result, we still don't know at which temperature a topologically close circular DNA molecule is finally denatured. The experiment of exposing such DNA at temperatures higher than 107°C (thus under pressure) and in the presence of high monovalent salt concentrations to prevent DNA breakage remains to be done.

10. Marguet and Forterre, 1994.

11. The original sentence was "I am willing to bet that the plectonemic coiling in your structure is radically wrong." A plectonemic coiling corresponds to the wrapping of a double helix in which the two strands are intertwined: that is, they may not be separated without uncoiling.

12. Watson and Crick, 1953b.

13. Wang, 2009; Forterre, 2012.

14. The number of topological links (linking number) of a covalently closed circular DNA double-stranded molecule is approximately equal to the sum of the twist (T) and writhing (W) numbers (L = T + W) (see Wang, 2009). T and W are complex values in the real 3D space; for simplification, they are roughly equivalent to the numbers of turns and superturns in a 2D space (Forterre, 2011). The linking number is a topological constant that cannot be changed as long as the two interlaced DNA strands remain intact.

15. In 1971, James Wang discovered the first DNA topoisomerase (a type I) in *Escherichia coli*, at that time called the omega protein (see Wang, 1971).

16. Type II topoisomerases can relax (or supercoil) DNA by catalyzing the crossing of two segments of the same double-stranded molecules, but they can also separate two distinct circular DNA molecules, such as two intertwined chromosomes at the end of bacterial DNA replication.

17. For a review of antibiotics inhibiting bacterial DNA gyrase, see Collin et al., 2011.

18. For a review of antitumoral drugs and antibiotics inhibiting DNA topoisomerases, see Pommier et al., 2010.

19. Guipaud et al., 1997.

20. López-García et al., 2000.

21. Marguet and Forterre, 1994.

22. At high temperatures, the thermal degradation of DNA starts even before the appearance of these breaks. The most fragile of the "strong" bonds found in our double helix is the one that binds A or G bases (called purine) to the rest of the nucleotide that carries them (more precisely to sugar, the deoxyribose). If one heats a DNA molecule, the A and G bases gradually detach

from the double helix: this is called depurination. At the place of the base that disappeared, the bond with the neighboring nucleotide is made fragile; it breaks in turn, which this time leads to the rupture of the DNA strand.

23. Peak et al., 1995.

24. The positive supercoiling by reverse gyrase cannot be a useful trick, since we observed that thermal degradation occurs at the same speed, whether the DNA is positively or negatively supercoiled (Marguet and Forterre, 1994).

25. We don't really know why, but the fact remains: the positively charged atoms of sodium or potassium (they are ions) protect the strong bonds that maintain the nucleotide skeleton of the DNA molecule from the effects of temperature. Marguet and Forterre, 1994, 1998.

26. Kopylov et al., 1993.

27. DiRuggiero et al., 1997; Gérard et al., 2001.

28. Jolivet et al., 2003.

29. Shin et al., 2014; Rouillon and White, 2011.

30. Zivanovic et al., 2009.

31. Llanos et al., 2000.

32. Lagorce et al., 2012.

33. Gorlas et al., 2015.

34. This is a simplified version of what really happens. In fact, there are always three molecules of tRNA bonded to a ribosome synthesizing proteins.

35. These enzymes are called amino-acyl tRNA synthetases.

36. Kowalak et al., 1994.

37. Edmonds et al., 1991; Noon et al., 2003.

38. Kowalak et al., 1994.

39. The carbon of the ribose and deoxyribose are labeled 1', 2', and so on to avoid confusion with the label of the carbon of the bases labeled 1, 2, and so on.

40. Eigner et al., 1961; Ginoza et al., 1964.

41. McCloskey et al., 2001; Noon et al., 2003.

42. Basak and Ghosh, 2005.

43. Pedersen and Reeh, 1978; Bini et al., 2002.

44. To my knowledge there are few studies that address the problem of the time required for the RNA message to move from the site of its synthesis in the nucleus to ribosomes in the cytoplasm.

45. Forterre, 1995.

46. Recall that protons are hydrogen atoms that have lost their single electron and thus carry a positive electric charge: H+.

47. Mulkidjanian et al., 2012.

48. Klein et al., 2004.

49. For an exhaustive look at the role of membranes in ion transport, see Konings, 2006.

50. In eukaryote cells, these reactions are produced in the internal membrane of specialized cellular compartments: mitochondria.

51. Mitchell, 1961.

52. Mulkidjanian et al., 2007.

53. Depending on the organisms, the ATP synthases use either the flow of protons or the flow of sodium to function. The membranes always being more permeable to protons than to sodium ions, which are much bigger, some hyperthermophiles prefer using sodium ATP synthases.

However, many microbes from hell, in particular those that live in the presence of oxygen, successfully use proton ATP synthases, thanks to their impermeable membrane. Proton ATP synthases enable the recovery of more energy in the presence of oxygen.

54. Van de Vossenberg et al., 1995. For a general review of the role of lipids in the thermoadaptation in archaea and bacteria, see Könings et al., 2002.

55. Harold and Van Brunt, 1977.

56. Curiously, while writing this book, I realized that no one had yet tested this hypothesis.

57. Sehgal et al., 1962.

58. Langworthy et al., 1972.

59. Langworthy et al., 1974.

60. Tornabene and Langworthy, 1979.

61. Sprott et al., 1991.

62. For a brief study of lipids from hyperthermophiles, see Koga, 2012.

63. Klein et al., 1979.

64. Damste et al., 2007.

65. Van de Vossenberg et al., 1995.

66. Lombard et al., 2012.

Chapter Four

1. The Crafoord Prize, established in 1980 by Anna-Greta and Holger Crafoord from Lund, is awarded by the Royal Swedish Academy of Sciences annually on a rotating basis between astronomy and mathematics, biosciences, and geosciences.

2. Some authors have derided the protein-based universal trees of life as "trees of 1%" because they are based on a small number of universal proteins (thirty to forty), that is, around 1% of proteins encoded by a typical bacterial genome (Dagan and Martin, 2006). They suggested replacing trees by networks that emphasize the importance of gene transfers in microbial evolution. However, in a Darwinian framework, gene transfers, as well as endosymbiosis, can be considered simply as a particular type of variation. In my opinion, these authors confuse gene and species trees. It's not a problem (it's even very nice) if a "good" species tree can be reconstructed from a very small number of genes. For reviews, see Gribaldo and Brochier, 2009; Forterre, 2010, 2012.

3. Woese, 1981.

4. The first two methods for sequencing DNA were designed independently in 1977 by Sanger and Gilbert. The Sanger method has for decades been the cornerstone of DNA sequencing methods used in laboratories and sequencing centers around the world.

5. For the original papers in which Woese and Fox published their discoveries, see Woese and Fox, 1977a; Fox et al., 1980; Woese, 1987; for a historical account see Sapp, 2009.

6. More recent methods rely on statistical methods such as maximum likelihood or Bayesian inference, which allow a choice between different trees, the most obvious ones depending on various evolutionary models. However, the best methods can fail if there is no phylogenetic signal or too much noise in a sequence alignment.

7. Schmidt et al., 1991; DeLong, 1992; Barns et al., 1994. The archaea detected in cold environments by microbial ecologists were first classified as Crenarchaeota. It was later found that most of them are members of a third major archaeal kingdom, the Thaumarchaeota; see chapter 5 and Brochier-Armanet et al., 2008a.

8. Karner et al., 2001.

9. Könnecke et al., 2005. In this paper, the authors reveal novel types of deep-branching archaeal rRNA, suggesting a new phylum, the Korarchaeota (from the Greek *kouros*, meaning "young men"). There is presently only one species of this group whose genome has been sequenced following arduous work in Stetter's laboratory, *Candidatus Korarchaeum cryptofilum* (Elkins et al., 2008).

10. Probst et al., 2013. These thaumarchaea were detected on the torsos of ten men.

11. Yates, 2007.

12. For a review of metagenomics analyses of extreme environments, see Cowan et al., 2015. We will see in chapter 5 the problems raised by genome reconstruction via metagenomic analysis.

13. Pace et al., 1986.

14. The two elongation factors help the ribosome to move along the messenger RNA during translation, using GTP instead of ATP as energy cofactor.

15. Iwabe et al., 1989.

16. In the case of eukaryotes, the proteins related to the ATP synthases of archaea and bacteria do not manufacture ATP, but degrade it. These ATPases are called vacuolary because they reside in the membranes of vacuoles. This is why we need mitochondria and their bacterial ATP synthases to breathe. We will return to this curious observation in chapter 5.

17. Gogarten et al., 1989.

18. Woese et al., 1990.

19. Stetter, 1996.

20. The phenotype of an ancestral organism at the base of a branch on a phylogenetic tree can be very different from the phenotype of the modern organism at the tip of the branch.

21. Miller and Lazcano, 1995.

22. Wächtershäuser, 1988; Huber and Wächtershäuser, 2006.

23. Keller et al., 1994. This collaboration ended rapidly after only two publications.

24. Martin and Russel, 2007; Martin et al., 2008; Russell et al., 2014.

25. In a recent paper, Russell's reactor produced very small RNA composed of two to four monomers (ribonucleotides); Burcar et al., 2015. For a general review of the origin of life problem, see Forterre and Gribaldo, 2007.

26. Arrhenius et al., 1999.

27. For Gilbert (1986), the RNA world preceded the cellular world; it was a world of "free-living" RNA molecules competing among each other. For most scientists, the RNA world now means a world of cells with RNA genomes (Forterre, 2005).

28. Woese, 1967; Crick, 1962; Orgel, 1968.

29. Lindhal, 1967. In contrast, RNA is stabilized at high temperatures by high concentrations of monovalent ions such as K+ (Hethke et al., 1999). However, K+ cannot replace Mg++ for ribozyme activity (Grosshans and Cech, 1989).

30. Forterre, 1995, 1996; Forterre et al., 1995.

31. Gogarten-Boekels et al., 1995.

32. Gomes et al., 2005. For a recent paper see Bottke et al., 2012.

33. The temperature of the early ocean is a controversial topic in the geosciences. For a hot ocean, see Robert and Chaussidon, 2006; for a milder ocean, see Blake et al., 2010.

34. Forterre and Philippe, 1999.

35. Willi Hennig (1913–76), a German entomologist, is considered the father of the cladis-

tics method (Hennig, 1966). A shared derived characteristic useful to group some organisms together is called a synapomorphy.

36. Forterre et al., 1992; see text at http://archaea.u-psud.fr/Forterre-Universaltree.pdf.

37. Forterre and Philippe, 1999.

38. Forterre, 1995; see text at http://archaea.u-psud.fr/1995_PForterre_Thermoreduction.pdf. A preliminary version of this hypothesis was also published in 1991 as a book chapter; see text at http://archaea.u-psud.fr/Forterre-Virus92.pdf/.

39. Glansdorff, 1999; Poole et al., 1999.

40. Galtier et al., 1999.

41. Boussau et al., 2008.

42. Groussin and Gouy, 2011. These three papers are rarely quoted by the community of scientists studying hyperthermophiles who usually really like the idea of a hyperthermophile LUCA.

43. One can easily imagine a cold LUCA with archaeal lipids, since the latter can function equally well from 0 to 110°C. For a review discussing the origin and evolution of lipids, as well as the great lipid divide, see Pereto et al., 2004; Glansdorff and Labedan, 2008; Lombard et al., 2012. The latter authors support the idea of a LUCA with both archaeal and bacterial lipids. The scenario of LUCA with bacterial/eukaryal lipids is briefly discussed in Forterre, 2013b.

44. Woese and Fox, 1997b.

45. Lazcano et al., 1988.

46. Mushegian and Koonin, 1996.

47. To explain the great DNA replication divide, Koonin and his colleagues then suggested that LUCA was indeed a DNA organism, but that this DNA was not replicated directly from another DNA molecule, but retrotranscribed from an RNA intermediate (Leipe et al., 1999). This type of reaction is performed by retroviruses, such as the AIDS virus, which transform their RNA genome into DNA that can then be integrated into the DNA genome of their cellular "host."

48. Koonin and Martin, 2005.

49. Denison et al., 2011; for other arguments in favor of an RNA-based LUCA, see Poole and Logan, 2005.

50. The ribonucleotide reductase (RNR) that removes the 2′ oxygen of the ribose and the thymidylate synthases that transform the base uridine into thymidine are sophisticated protein enzymes that use complex chemistry. In particular, the reaction catalyzed by the RNR involves very reactive radical groups, suggesting that this reaction could not have been catalyzed by RNA ribozymes because these groups would have indicated that DNA most likely originated after proteins (Freeland et al., 1999).

51. For a time I thought that the root between eukaryotes and a common ancestor of archaea and bacteria was necessary to justify the existence in LUCA of eukaryote characteristics that could have been very ancient, such as genes in pieces. But a colleague from New Zealand, Anthony Poole, pointed out to me that these characteristics could just as well have been lost independently in archaea and bacteria and that the "bacterial rooting" did not prevent imagining the existence of ancestral eukaryotic characteristics in LUCA.

52. Forterre, 2002a.

53. Forterre, 2006a. I now think that two transfer events of DNA from viruses to cells are more likely, considering that DNA replication proteins in archaea and eukarya are very similar. However, because these proteins are also present in several families of large DNA viruses, such

as Mimivirus, it is difficult to obtain a clear answer. It is possible that several eukaryotic DNA replication proteins were introduced in the eukaryotic lineage from different viruses.

54. Liu et al., 1979. The T4 type II DNA topoisomerase has no gyrase activity. It branches at midpoint between bacterial and eukaryotic type II DNA topoisomerases in phylogenetic trees of these enzymes (Forterre and Gadelle, 2009).

55. Forterre, 1999.

56. For the most recent study on unusual base modifications in bacterial DNA viruses, see Warren, 1980. This very important topic is obviously no longer in vogue.

57. Forterre and Gadelle, 2009.

58. DNA polymerases involved in DNA replication (replicase) are associated with several accessory proteins that increase the accuracy and speed of DNA replication (processivity).

59. Forterre, 2002a, 2005, 2006b.

60. For a hypothetical scenario of the transfer of DNA from viruses to cells, see Forterre, 2005.

61. Whitfield, 2006; Zimmer, 2006.

62. For a review minimizing the role of viruses in cellular evolution and promoting the traditional view on viruses, see Moreira and López-García, 2009. For critics of some of their arguments, see Forterre, 2010; Forterre and Prangishvili, 2009, 2013.

63. New genes in cells or viruses can originate through gene duplication, recombination, various polymerase mistakes, or positive selection of fortuitous translation products of randomly synthesized RNA sequences (protogenes) (Carvunis et al., 2012). The *de novo* origin of new genes overlapping old ones in the noncoding strand of RNA viruses has been well described by Karlin and his colleagues (Sabath et al., 2012).

64. Benson et al., 1999, 2004.

65. Lwoff, 1953; Temin, 1971. The escape theory derived from the discovery of prophages in bacterial genomes and proviruses in eukaryotic genomes. It was assumed that such elements were at the origin of viruses.

66. Besides Adenovirus, the adenovirus/ PRD1 lineage also includes most eukaryotic viruses with large DNA genomes related to Mimivirus (Abrescia et al., 2012). Except Adenovirus, the genomes of all viruses belonging to this lineage are always enclosed in an internal lipid vesicle within the capsid. They also share the same type of protein for genome packaging, a so-called AAA+ ATPase. However, they can use very different types of DNA replication proteins, indicating that viruses can exchange this type of protein during their evolution.

67. Baker et al., 2005; Abrescia et al., 2012; Pietilä et al., 2013.

68. Viruses of the HK97 lineage also share a packaging ATPase that is not homologous to those of the PRD1/Adenovirus lineage.

69. The origin of viruses can be associated with the invention of virions as a novel powerful mechanism to disseminate genetic information. Virions may have appeared from the cellular structure of the RNA cell (membrane vesicles, chromosome, proteic compartment) that acquire the ability to encapsulate small RNA and carry them from one cell to another; see Forterre and Krupovic, 2012.

70. In scientific terms, viruses are thus polyphyletic.

71. For an exhaustive look at viral lineages, see the review by Abrescia et al., 2012.

72. Viruses have been defined as capsid-encoding organisms by Raoult and Forterre (2008). However, virologists also give the name "virus" to infectious RNA responsible for some plant

diseases. In that case, the viral genome is somehow confused with the virion. In my opinion, this is misleading, since infectious plasmids in archaea and bacteria are not called viruses.

73. Forterre, 2006b.

74. Bamford, 2003.

75. For a short and incisive paper on this topic, see Krupovic and Bamford, 2010.

76. Raoult et al., 2004; Colson et al., 2012; see Forterre 2010 for a critical study of giant viruses.

77. Forterre et al., 2014.

78. Quax et al., 2011.

79. Forterre and Prangishvili, 2009, 2013.

80. "The Scholar and the Horse: A Modern Fable" (January 23, 2013): http://sciences.blogs .liberation.fr/home/2013/01/ourasi-versus-woese-in-memoriam-html/.

81. For a scientific review in honor of Woese and Zillig, two main characters of this book, see Albers et al., 2014.

Chapter Five

1. Kornberg, 1991.

2. Saiki et al., 1988.

3. Fleischmann et al., 1995.

4. Bult et al., 1996.

5. For early reviews on archaeal genomes, see Olsen and Woese, 1997; Forterre, 1997.

6. She et al., 2001.

7. Cohen et al., 2003.

8. Myllykallio et al., 2000; Matsunaga et al., 2001.

9. Myllykallio et al., 2002.

10. Kuhn et al., 2002.

11. Myllykallio et al., 2000, 2002.

12. Skouloubris et al., 2015.

13. Cossu et al., 2015.

14. With a few exceptions, some universal proteins can be missing in parasitic organisms with very reduced genomes.

15. Koonin, 2003; Delaye et al., 2005.

16. This is the case of the Hepatitis delta virus: Lai, 2005.

17. Lecompte et al., 2002.

18. Woese and Fox, 1977b.

19. Kae1 and Sua5 were originally named respectively YgjD and YrdC; in bacteria, the mito-chondrial version of Kae1/YgjD is called Qri7. The human Kae1 protein is also called OSGEP. A new unifying nomenclature has recently been proposed (Thiaville et al., 2014a).

20. For a review, see Thiaville et al., 2014a.

21. Perrochia initially succeeded in reproducing the synthesis of t6A in a test tube (*in vitro*) by mixing a transfer RNA with threonine, ATP, and carbamate in the presence of universal pro-teins and accessory proteins of bacteria, archaea, or eukaryotes (Perrochia et al., 2013). He and a Canadian group then succeeded in reproducing this reaction by using only the two universal proteins used by the mitochondria, one of eukaryote origin, the other of bacterial origin (Wan et al., 2013; Thiaville et al., 2014b).

22. Mushegian and Koonin, 1996.

23. Mulkidjanian et al., 2007.

24. Koonin and Martin, 2005.

25. Peretó et al., 2004.

26. Woese, 1998; for critics of the communal LUCA, see Poole, 2009; Forterre, 2012.

27. Forterre, 2015.

28. Prangishvili et al., 2006.

29. Quemin et al., 2013.

30. Krupovic et al., 2014.

31. Krupovic et al., 2013; Koonin and Krupovic, 2015.

32. For a review of these fantastic new tools obtained thanks to fundamental research, see Sternberg and Doudna, 2015.

33. Mulkidjanian et al., 2012.

34. Dibrova et al., 2015.

35. Pino et al., 2015.

36. Ferus et al., 2015.

37. Fallah-Araghi et al., 2014. For a general review of the origin of life problem, see Forterre and Gribaldo, 2007.

38. Forterre, 2002b. There are more than two hundred reverse gyrase genes in the database in 2015 versus seventeen in 2002; all are encoded by genomes of thermophiles or hyperthermophiles. Not a single reverse gyrase gene has been found in the thousands of genomes from mesophiles now available for analysis.

39. Confalonieri et al., 1993.

40. Forterre et al., 1995.

41. Stetter suggested that some simpler device could have replaced reverse gyrase in primitive hyperthermophiles. Why not? Once again, we would like to know more about the role of reverse gyrase *in vivo* to evaluate whether this is a realistic hypothesis.

42. Forterre et al., 2000.

43. Brochier-Armanet and Forterre, 2007.

44. When complete genome sequences became available, phylogenies based on ribosomal RNA were gradually replaced by those of ribosomal proteins. The trees thus obtained are *a priori* more reliable than those obtained with 16S ribosomal RNA because they enable the comparison of the sequences of twenty letters (the twenty amino acids) rather than of four letters (the nucleotides). However, the number of amino acids in ribosomal protein is much smaller than the number of nucleotides in 16S ribosomal RNA. To obtain enough informative positions, one must thus add end to end the sequences of amino acids of many ribosomal proteins. This process is called concatenation.

45. Groussin and Gouy, 2011.

46. Atomi et al., 2004.

47. Zhang et al., 2013.

48. Bergerat et al., 1994.

49. Through meiosis, each individual manufactures sex cells whose genome is a mosaic of that of his or her parents. The chromosomes of a baby born from the fertilization of an ovum by a spermatozoid are thus an original combination of those of the parents: each individual issued from sexual reproduction is unique.

50. De Massy et al., 1995.

51. Bergerat et al., 1997.

52. Hartung and Puchta, 2001.

53. Sugimoto-Shirasu et al., 2002.

54. Yoon et al., 2014.

55. Marc et al., 2002.

56. Lindas et al., 2008; Samson et al., 2008.

57. Cox et al., 2008.

58. The grouping of eukaryotes and Crenarchaeota was proposed in the 1980s by James Lake, who called them eocytes ("cells of dawn") (Lake et al., 1984). He originally based his proposal on ribosome structure observed through electron microscopy, then on ribosomal RNA phylogeny, using his homemade tree-building method. For a detailed description of the eocyte story, see Sapp, 2009.

59. This point is discussed with comparative figures in Forterre, 2013b.

60. Woese justified his own "gradism" by suggesting that evolution from LUCA to the three domains did not follow the classical rules of Darwinian evolution, with speciation, but was "communal," with all organisms continuously exchanging characters via multiple lateral gene transfers (Woese). Cladistics rules thus did not apply before a "Darwinian threshold" that was crossed by ancestors of archaea and eukaryotes only after their divergence. For criticism of this view, see Poole, 2009; Forterre, 2012.

61. Forterre, 2015.

62. Forterre, 1992.

63. Forterre, 2013b.

64. For discussion of the term "prokaryote," see Pace, 2009.

65. Penny et al., 2009.

66. Lane et al., 2007.

67. Valentine (2007) suggested that this explains the success of archaea in energy- and/or nutrient-poor environments.

68. Forterre, 1995. Young and colleagues described a new group of eukaryotic-like RNA viruses in a Yellowstone hot spring enriched in archaea (Bolduc et al., 2012). However, up to now, they have not succeeded in isolating one of these viruses or in demonstrating that they indeed infect an archaeon; let's wait and see.

69. Woese, 2002.

70. For papers from supporters of fusion hypotheses explaining the emergence of specific eukaryotic features, see López-García and Moreira, 2006; Martin and Koonin, 2006; Lane and Martin, 2010; Koonin et al., 2015; and references therein.

71. For a description of the various fusion hypotheses proposed during the last thirty years, see Forterre, 2011b. For critics of fusion hypotheses, see De Duve, 2007; Kurland et al., 2006; Forterre, 2011b, 2013b.

72. Koonin, 2010a.

73. Koonin, 2010a.

74. El Albani et al., 2010.

75. See, for instance, Martijn and Ettema, 2013.

76. Rinke et al., 2013, have described many new archaeal "phylums," including Woesearchaeota and Pacearchaeota, based on metagenomic data. These correspond to small and rapidly evolving archaea that probably form a deep branching group of Euryarchaeota (Forterre, unpublished analyses).

77. *Cenarchaeum symbiosum* is a marine archaea that lives in symbiosis in sponges. Its genome was obtained without needing to culture it, thanks to DNA amplification techniques made possible by the DNA polymerase of microbes from hell.

78. Brochier-Armanet et al., 2008a. The name "Thaumarchaeota" was suggested to us by Guennadi Sezonov, a professor at Université Pierre et Marie Curie in Paris who at the time was working with us on viruses at the Institut Pasteur.

79. Pester et al., 2011; Brochier-Armanet et al., 2012.

80. Nunoura et al., 2011.

81. They assumed that the Greek word *aiga* meant dawn, as they were supposed to be primitive cells. In fact, a Greek-speaking colleague told me that the correct word for dawn in Greek is *auge*, and that *aigi* means "goat" in Greek.

82. Brochier-Armanet et al., 2008b.

83. Spang et al., 2012.

84. Brochier-Armanet et al., 2008a.

85. Koonin and Yutin, 2014.

86. Koonin, 2010a, 2010b.

87. Forterre, 2013b.

88. Guy and Ettema, 2011.

89. For a paper supporting these analyses, see Williams et al., 2013.

90. Gribaldo et al., 2010.

91. Spang et al., 2015.

92. Raymann et al., 2015.

93. An analysis of Loki 16S ribosomal RNA sequences present in the Arctic sample indicated that Loki is a member of a large group of uncultivated archaea known for a long time by environmental microbiologists and named the Deep Sea Archaeal Group (DSAG) (Jorgensen et al., 2013).

94. Brochier et al., 2004, 2005.

95. Kurland et al., 2006.

96. A change in the tempo of evolution.

97. Forterre, 2006a.

References

Abrescia, N. G., Bamford, D. H., Grimes, J. M., and Stuart, D. I. 2012. Structure unifies the viral universe. *Annual Review of Biochemistry* 81:795–822.

Alain, K., Marteinsson, V. T., Miroshnichenko, M. L., Bonch-Osmolovskaya, E. A., Prieur, D., Birrien, J., et al. 2002. Marinitoga piezophila sp. nov., a rod-shaped, thermo-piezophilic bacterium isolated under high hydrostatic pressure from a deep-sea hydrothermal vent. *International Journal of Systematic and Evolutionary Microbiology* 52(Pt. 4):1331–39.

Albers, S. V., Forterre, P., Prangishvili, D., and Schleper, C. 2013. The legacy of Carl Woese and Wolfram Zillig: From phylogeny to landmark discoveries. *Nature Reviews Microbiology* 11:713–19.

Arnold, H. P., Zillig, W., Ziese, U., Holz, I., Crosby, M., Utterback, T., et al. 2000. A novel lipothrixvirus, SIFV, of the extremely thermophilic crenarchaeon Sulfolobus. *Virology* 267: 252–66.

Arrhenius, G., Bada, J. L., Joyce, G. F., Lazcano, A., Miller, S., and Orgel, L. E. 1999. Origin and ancestor: Separate environments. *Science* 283:792.

Atomi, H., Matsumi, R., and Imanaka, T. 2004. Reverse gyrase is not a prerequisite for hyperthermophilic life. *Journal of Bacteriology* 186:4829–33.

Baker, M. L., Jiang, W., Rixon, F. J., and Chiu, W. 2005. Common ancestry of herpes viruses and tailed DNA bacteriophages. *Journal of Virology* 79:14967–70.

Bamford, D. H. 2003. Do viruses form lineages across different domains of life? *Research in Microbiology* 154:231–36.

Barns, S. M., Fundyga, R. E., Jeffries, M. W., and Pace, N. R. 1994. Remarkable archaeal diversity detected in a Yellowstone National Park hot spring environment. *Proceedings of the National Academy of Sciences USA* 91:1609–13.

Baross, J. A., and Deming, J. W. 2003. Growth of "black smoker" bacteria at temperatures of at least 250°C. *Nature* 303:423.

Basak, S., and Ghosh, T. C. 2005. On the origin of genomic adaptation at high temperature for prokaryotic organisms. *Biochemical and Biophysical Research Communications* 330:629–32.

Basen, M., Schut, G. J., Nguyen, D. M., Lipscomb, G. L., Benn, R. A., Prybol, C. J., et al. 2014. Single gene insertion drives bioalcohol production by a thermophilic archaeon. *Proceedings of the National Academy of Sciences USA* 111:17618–23.

Benson, S. D., Bamford, J. K., Bamford, D. H., and Burnett, R. M. 1999. Viral evolution revealed by bacteriophage PRD1 and human adenovirus coat protein structures. *Cell* 98:825–33.

———. 2004. Does common architecture reveal a viral lineage spanning all three domains of life? *Molecular Cell* 16:673–85.

Bergerat, A., Gadelle, D., and Forterre, P. 1994. Purification of a DNA topoisomerase II from the hyperthermophilic archaeon Sulfolobus shibatae. A thermostable enzyme with both bacterial and eucaryal features. *Journal of Biological Chemistry* 269:27663–69.

Bergerat, A., de Massy, B., Gadelle, D., Varoutas, P. C., Nicolas, A., and Forterre, P. 1997. An atypical topoisomerase II from Archaea with implications for meiotic recombination. *Nature* 386:414–17.

Bettstetter, M., Peng, X., Garrett, R. A., and Prangishvili, D. 2003. AFV1, a novel virus infecting hyperthermophilic archaea of the genus acidianus. *Virology* 315:68–79.

Beveridge, T. J. 1999. Structures of gram-negative cell walls and their derived membrane vesicles. *Journal of Bacteriology* 181:4725–33.

Bini, E., Dikshit, V., Dirksen, K., Drozda, M., and Blum, P. 2002. Stability of mRNA in the hyperthermophilic archaeon Sulfolobus solfataricus. *RNA* 8:1129–36.

Bize, A., Karlsson, E. A., Ekefjärd, K., Quax, T. E., Pina, M., Prevost, M. C., et al. 2009. A unique virus release mechanism in the Archaea. *Proceedings of the National Academy of Sciences USA* 106:11306–11.

Blake, R. E., Chang, S. J., and Lepland, A. 2010. Phosphate oxygen isotopic evidence for a temperate and biologically active Archaean ocean. *Nature* 464:1029–32.

Blöchl, E., Rachel, R., Burggraf, S., Hafenbradl, D., Jannasch, H. W., and Stetter, K. O. 1997. Pyrolobus fumarii, gen. and sp. nov., represents a novel group of archaea, extending the upper temperature limit for life to 113 degrees C. *Extremophiles* 1:14–21.

Bolduc, B., Shaughnessy, D. P., Wolf, Y. I., Koonin, E. V., Roberto, F. F., and Young, M. 2012. Identification of novel positive-strand RNA viruses by metagenomic analysis of archaea-dominated Yellowstone hot springs. *Journal of Virology* 86:5562–73.

Bottke, W. F., Vokrouhlický, D., Minton, D., Nesvorný, D., Morbidelli, A., Brasser, R., et al. 2012. An Archaean heavy bombardment from a destabilized extension of the asteroid belt. *Nature* 485:78–81.

Boussau, B., Blanquart, S., Necsulea, A., Lartillot, N., and Gouy, M. 2008. Parallel adaptations to high temperatures in the Archaean eon. *Nature* 456:942–45.

Brochier, C., Forterre, P., and Gribaldo, S. 2004. Archaeal phylogeny based on proteins of the transcription and translation machineries: Tackling the Methanopyrus kandleri paradox. *Genome Biology* 5(3):R17.

Brochier, C., Gribaldo, S., Zivanovic, Y., Confalonieri, F., and Forterre, P. 2005. Nanoarchaea: representatives of a novel archaeal phylum or a fast-evolving euryarchaeal lineage related to Thermococcales? *Genome Biology* 6(5):R42.

Brochier-Armanet, C., Boussau, B., Gribaldo, S., and Forterre, P. 2008a. Mesophilic Crenarchaeota: Proposal for a third archaeal phylum, the Thaumarchaeota. *Nature Reviews Microbiology* 3:245–52.

Brochier-Armanet, C., and Forterre, P. 2007. Widespread distribution of archaeal reverse gyrase in thermophilic bacteria suggests a complex history of vertical inheritance and lateral gene transfers. *Archaea* 2:83–93.

Brochier-Armanet, C., Gribaldo, S., and Forterre, P. 2008b. A DNA topoisomerase IB in

Thaumarchaeota testifies for the presence of this enzyme in the last common ancestor of Archaea and Eucarya. *Biology Direct* 3:54. doi:10.1186/1745-6150-3-54.

———. 2011. Phylogeny and evolution of the Archaea: One hundred genomes later. *Current Opinion in Microbiology* 14:274–81.

———. 2012. Spotlight on the Thaumarchaeota. *International Journal of Systematic and Evolutionary Microbiology* 6:227–30.

Brock, T. D. 1995. The road to Yellowstone—and beyond. *Annual Review of Microbiology* 49:1–28.

Brock, T. D., Brock, K. M., Belly, R. T., and Weiss, R. L. 1972. Sulfolobus: A new genus of sulfur-oxidizing bacteria living at low pH and high temperature. *Archiv für Mikrobiologie* 84: 54–68.

Brock, T. D., and Freeze, H. 1969. Thermus aquaticus gen. n. and sp. n., a nonsporulating extreme thermophile. *Journal of Bacteriology* 98:289–97.

Bult, C. J., White, O., Olsen, G. J., Zhou, L., Fleischmann, R. D., Sutton, G. G., et al. 1996. Complete genome sequence of the methanogenic archaeon Methanococcus jannaschii. *Science* 273:1058–73.

Burcar, B. T., Barge, L. M., Trail, D., Watson, E. B., Russell, M. J., and McGown, L. B. 2015. RNA oligomerization in laboratory analogues of alkaline hydrothermal vent systems. *Astrobiology* 15:509–22.

Cacace, M. G., De Rosa, M., and Gambacorta, A. 1976. DNA-dependent RNA polymerase from the thermophilic bacterium Caldariella acidophila: Purification and basic properties of the enzyme. *Biochemistry* 15:1692–96.

Carvunis, A. R., Rolland, T., Wapinski, I., Calderwood, M. A , Yildirim M. A., Simonis, N., et al. 2012. Proto-genes and de novo gene birth. *Nature* 487:370–74.

Cech, T. R. 2000. The ribosome is a ribozyme. *Science* 289:878–79.

Cech, T. R., and Bass, B. L. 1986. Biological catalysis by RNA. *Annual Review of Biochemistry* 55:599–629.

Cohen, G. N., Barbe, V., Flament, D., Galperin, M., Heilig, R., Lecompte, O., et al. 2003. An integrated analysis of the genome of the hyperthermophilic archaeon Pyrococcus abyssi. *Molecular Microbiology* 47:1495–512.

Collin, F., Karkare, S., and Maxwell, A. 2011. Exploiting bacterial DNA gyrase as a drug target: Current state and perspectives. *Applied Microbiology and Biotechnology* 92:479–97.

Colson, P., de Lamballerie, X., Fournous, G., and Raoult, D. 2012. Reclassification of giant viruses composing a fourth domain of life in the new order Megavirales. *Intervirology* 55:321–32.

Confalonieri, F., Elie, C., Nadal, M., de la Tour, C., Forterre, P., and Duguet, M. 1993. Reverse gyrase: A helicase-like domain and a type I topoisomerase in the same polypeptide. *Proceedings of the National Academy of Sciences USA* 90:4753–57.

Corliss, J. B., Baross, J. A., and Hoffman, S. E. 1981. An hypothesis concerning the relationship between submarine hot springs and the origin of life on earth. *Oceanologica acta*. In Proceedings of the 26th International Geological Congress, Geology of Oceans Symposium, Paris, July 7–17, 1980, 59–69.

Corliss, J. B., Dymond, J., Gordon, L. I., Edmond, J. M., von Herzen R. P., Ballard R. D., et al. 1979. Submarine thermal springs on the Galapagos rift. *Science* 203:1073–83.

Cortez, D., Forterre, P., and Gribaldo, S. 2009. A hidden reservoir of integrative elements is the major source of recently acquired foreign genes and ORFans in archaeal and bacterial genomes. *Genome Biology* 10:R65.

Cossu, M., Da Cunha, V., Toffano-Nioche, C., Forterre, P., and Oberto, J. 2015. Comparative genomics reveals conserved positioning of essential genomic clusters in highly rearranged Thermococcales chromosomes. *Biochimie* 118:313–21.

Cowan, D. A. 2004. The upper temperature for life: Where do we draw the line? *Trends in Microbiology* 12:58–60.

Cowan, D. A., Ramond, J. B., Makhalanyane, T. P., and De Maayer, P. 2015. Metagenomics of extreme environments. *Current Opinion in Microbiology* 25:97–102.

Cox, C. J., Foster, P. G., Hirt, R. P., Harris, S. R., and Embley, T. M. 2008. The archaebacterial origin of eukaryotes. *Proceedings of the National Academy of Sciences USA* 105:20356–61.

Crick, F. H. 1962. The genetic code. *Scientific American* 207:66–74.

Dagan, T., and Martin, W. 2006. The tree of one percent. *Genome Biology* 7:118.

Damsté, J. S., Rijpstra, W. I., Hopmans, E. C., Schouten, S., Balk, M., and Stams, A. J. 2007. Structural characterization of diabolic acid-based tetraester, tetraether and mixed ether/ester, membrane-spanning lipids of bacteria from the order Thermotogales. *Archives of Microbiology* 188:629–41.

Darland, G., Brock, T. D., Samsonoff, W., and Conti, S. F. 1970. A thermophilic, acidophilic mycoplasma isolated from a coal refuse pile. *Science* 170:1416–18.

Davis, B. M. 1897. The vegetation of the hot springs of Yellowstone Park. *Science*, 145–57.

Deatherage, B. L., and Cookson, B. T. 2012. Membrane vesicle release in bacteria, eukaryotes, and archaea: A conserved yet underappreciated aspect of microbial life. *Infection and Immunity* 80:1948–57.

De Duve, C. 2007. The origin of eukaryotes: A reappraisal. *Nature Reviews Genetics* 8:395–403.

Delaye, L., Becerra, A., and Lazcano, A. 2005. The last common ancestor: What's in a name? *Origins of Life and Evolution of Biospheres* 35:537–54.

DeLong, E. F. 1992. Archaea in coastal marine environments. *Proceedings of the National Academy of Sciences USA* 89:5685–89.

De Massy, B., Rocco, V., and Nicolas, A. 1995. The nucleotide mapping of DNA double-strand breaks at the CYS3 initiation site of meiotic recombination in Saccharomyces cerevisiae. *EMBO Journal* 14:4589–98.

Denison, M. R., Graham, R. L., Donaldson, E. F., Eckerle, L. D., and Baric, R. S. 2011. Coronaviruses: An RNA proofreading machine regulates replication fidelity and diversity. *RNA Biology* 8:270–79.

Devos, D. P., Jogler, C., and Fuerst, J. A. 2013. The 1st EMBO workshop on PVC bacteria-Planctomycetes-Verrucomicrobia-Chlamydiae superphylum: Exceptions to the bacterial definition? *Antonie van Leeuwenhoek* 104:443–49.

Dibrova, D. V., Galperin, M. Y., Koonin, E. V., and Mulkidjanian, A. Y. 2015. Ancient systems of sodium/potassium homeostasis as predecessors of membrane bioenergetics. *Biochemistry (Moscow)* 80:495–516.

DiRuggiero, J., Santangelo, N., Nackerdien, Z., Ravel, J., and Robb, F. T. 1997. Repair of extensive ionizing-radiation DNA damage at 95 degrees C in the hyperthermophilic archaeon Pyrococcus furiosus. *Journal of Bacteriology* 179:4643–45.

Dorman, C. J. 2013. Co-operative roles for DNA supercoiling and nucleoid-associated proteins in the regulation of bacterial transcription. *Biochemical Society Transactions* 41:542–47.

Drlica, K. 1992. Control of bacterial DNA supercoiling. *Molecular Microbiology* 6:425–33.

Dubey, G. P., and Ben-Yehuda, S. 2011. Intercellular nanotubes mediate bacterial communication. *Cell* 144:590–600.

Dupressoir, A., Lavialle, C., and Heidmann, T. 2012. From ancestral infectious retroviruses to bona fide cellular genes: Role of the captured syncytins in placentation. *Placenta* 33:663–71.

Edmonds, C. G., Crain, P. F., Gupta, R., Hashizume, T., Hocart, C. H., Kowalak, J. A., et al. 1991. Posttranscriptional modification of tRNA in thermophilic archaea (Archaebacteria). *Journal of Bacteriology* 173:3138–48.

Eigner J., Boedtker, H., and Michaels G. 1961. The thermal degradation of nucleic acids. *Biochimica et Biophysica Acta* 51:165–68.

El Albani, A., Bengtson, S., Canfield, D. E., Bekker, A., Macchiarelli, R., Mazurier, A., et al. 2010. Large colonial organisms with coordinated growth in oxygenated environments 2.1 Gyr ago. *Nature* 466:100–104.

Elkins, J. G., Podar, M., Graham, D. E., Makarova, K. S., Wolf, Y., Randau, L., et al. 2008. A korarchaeal genome reveals insights into the evolution of the Archaea. *Proceedings of the National Academy of Sciences USA* 105:8102–7.

Elleuche, S., Schröder, C., Sahm, K., and Antranikian, G. 2014. Extremozymes-biocatalysts with unique properties from extremophilic microorganisms. *Current Opinion in Biotechnology* 29:116–23.

Fallah-Araghi, A., Meguellati, K., Baret, J. C., El Harrak, A., Mangeat, T., and Karplus, M. 2014. Enhanced chemical synthesis at soft interfaces: A universal reaction-adsorption mechanism in microcompartments. *Physical Review Letters* 112(2):028301.

Ferus, M., Nesvorný, D., Šponer, J., Kubelík, P., Michalčíková, R., Shestivská, V., et al. 2015. High-energy chemistry of formamide: a unified mechanism of nucleobase formation. *Proceedings of the National Academy of Sciences USA* 112:657–62.

Fiala, G., and Stetter, K. O. 1986. Pyrococcus furiosus sp. nov. represents a novel genus of marine heterotrophic archaebacteria growing optimally at 100°C. *Archives of Microbiology* 145:56–61.

Fleischmann, R. D., Adams, M. D., White, O., Clayton, R. A., Kirkness, E. F., Kerlavage, A. R., et al. 1995. Whole-genome random sequencing and assembly of Haemophilus influenzae Rd. *Science* 269:496–512.

Forterre, P. 1992. Neutral terms. *Nature* 335:305.

———. 1995. Thermoreduction: A hypothesis for the origin of prokaryotes. *Comptes Rendus de l'Académie des Sciences* 318(4): 415–22.

———. 1996. A hot topic: The origin of hyperthermophiles. *Cell* 85:789–92.

———. 1997. Archaea: What can we learn from their sequences? *Current Opinion in Genetics Development* 7:764–70.

———. 1999. Displacement of cellular proteins by functional analogues from plasmids or viruses could explain puzzling phylogenies of many DNA informational proteins. *Molecular Microbiology* 33:457–65.

———. 2002a. The origin of DNA genomes and DNA replication proteins. *Current Opinion in Microbiology* 5:525–32.

———. 2002b. A hot story from comparative genomics: Reverse gyrase is the only hyperthermophile-specific protein. *Trends in Genetics* 18:236–37.

———. 2005. The two ages of the RNA world, and the transition to the DNA world: A story of viruses and cells. *Biochimie* 87:793–803.

———. 2006a. Three RNA cells for ribosomal lineages and three DNA viruses to replicate their genomes: A hypothesis for the origin of cellular domain. *Proceedings of the National Academy of Sciences USA* 103:3669–74.

———. 2006b. The origin of viruses and their possible roles in major evolutionary transitions. *Virus Research* 117:5–16.

———. 2010. Giant viruses: Conflicts in revisiting the virus concept. *Intervirology* 53:362–78.

———. 2011a. A new fusion hypothesis for the origin of Eukarya: Better than previous ones, but probably also wrong. *Research in Microbiology* 162:77–91.

———. 2011b. Introduction and historical perspective. In *DNA Topoisomerases and Cancer.* Y. Pommier (ed.). Humana Press, Springer-Verlag, 1–52.

———. 2012. Darwin's goldmine is still open: Variation and selection run the world. *Frontiers in Cellular and Infection Microbiology* 2:106.

———. 2013a. The virocell concept and environmental microbiology. *International Journal of Systematic and Evolutionary Microbiology* 7:233–36.

———. 2013b. The common ancestor of archaea and eukarya was not an archaeon. *Archaea* 2013:372–96. doi:10.1155/2013/372396.

———. 2015. The universal tree of life: An update. *Frontiers in Microbiology* 6:717. doi:10.3389/fmicb.2015.00717.

Forterre, P., Benachenhou-Lahfa, N., Confalonieri, F., Duguet, M., Elie, C., and Labedan, B. 1992. The nature of the last universal ancestor and the root of the tree of life: Still open questions. *Biosystems* 28:15–32.

Forterre, P., Bouthier de la Tour, C., Philippe, H., and Duguet, M. 2000. Reverse gyrase from hyperthermophiles: Probable transfer of a thermoadaptation trait from archaea to bacteria. *Trends in Genetics* 16:152–54.

Forterre, P., Confalonieri, F., Charbonnier, F., and Duguet, M. 1995. Speculations on the origin of life and thermophily: Review of available information on reverse gyrase suggests that hyperthermophilic procaryotes are not so primitive. *Origins of Life and Evolution of Biospheres* 25:235–49.

Forterre, P., Elie, C., and Kohiyama, M. 1984. Aphidicolin inhibits growth and DNA synthesis in halophilic arachaebacteria. *Journal of Bacteriology* 159:800–802.

Forterre, P., and Gadelle, D. 2009a. Phylogenomics of DNA topoisomerases: Their origin and putative roles in the emergence of modern organisms. *Nucleic Acids Research* 37:679–92.

Forterre, P., and Gribaldo, S. 2007. The origin of modern terrestrial life. *The Human Frontier Science Program Journal* 1:156–68.

Forterre, P., and Krupovic, M. 2012. The origin of virions and virocells: The escape hypothesis revisited. In *Viruses: Essential Agents of Life.* Ed. G. Witzany. Springer Science + Business Media Dordrecht.

Forterre, P., Krupovic, M., and Prangishvili, D. 2014. Cellular domains and viral lineages. *Trends in Microbiology* 22:554–58.

Forterre, P., Mirambeau, G., Jaxel, C., Nadal, M., and Duguet, M. 1985. High positive supercoiling *in vitro* catalyzed by an ATP and polyethylene glycol-stimulated topoisomerase from Sulfolobus acidocaldarius. *EMBO Journal* 4:2123–28.

Forterre, P., and Philippe, H. 1999. Where is the root of the universal tree of life? *Bioessays* 21:871–79.

Forterre, P., and Prangishvili, D. 2009b. The great billion-year war between ribosome- and capsid-encoding organisms (cells and viruses) as the major source of evolutionary novelties. *Annals of the New York Academy of Sciences* 1178:65–77.

———. 2013. The major role of viruses in cellular evolution: Facts and hypotheses. *Current Opinion in Virology* 3:558–65.

Fox, G. E., Stackebrandt, E., Hespell, R. B., Gibson, J., Maniloff, J., Dyer, T. A., et al. 1980. The phylogeny of prokaryotes. *Science* 209:457–63.

Freeland, S. J., Knight, R. D., and Landweber, L. F. 1999. Do proteins predate DNA? *Science* 286:690–92.

Friend, T. 2007. *The third domain: The untold story of Archaea and the future of biotechnology.* Washington, DC: Joseph Henry Press.

Galtier, N., Tourasse, N., and Gouy, M. 1999. A nonhyperthermophilic common ancestor to extant life forms. *Science* 283:220–21.

Garrett, R., and Klenk, H. P. 2007. *Archaea.* Blackwell.

Gaudin, M., Gauliard, E., Schouten, S., Houel-Renault, L., Lenormand, P., Marguet, E., et al. 2013. Hyperthermophilic archaea produce membrane vesicles that can transfer DNA. *Environmental Microbiology Reports* 5:109–16.

Gaudin, M., Krupovic, M., Marguet, E., Gauliard, E., Cvirkaite-Krupovic, V., Le Cam, E., et al. 2014. Extracellular membrane vesicles harbouring viral genomes. *Environmental Microbiology* 16:1167–75.

Gellert, M., Mizuuchi, K., O'Dea, M. H., and Nash, H. A. 1976. DNA gyrase: An enzyme that introduces superhelical turns into DNA. *Proceedings of the National Academy of Sciences USA* 73:3872–76.

Gérard, E., Jolivet, E., Prieur, D., and Forterre, P. 2001. DNA protection mechanisms are not involved in the radioresistance of the hyperthermophilic archaea Pyrococcus abyssi and P. furiosus. *Molecular Genetics and Genomics* 266:72–78.

Geslin, C., Le Romancer, M., Erauso, G., Gaillard, M., Perrot, G., and Prieur, D. 2003. PAV1, the first virus-like particle isolated from a hyperthermophilic euryarchaeote, Pyrococcus abyssi. *Journal of Bacteriology* 185:3888–94.

Gilbert, W. 1978. Why genes in pieces? *Nature* 271:501.

———. 1986. Origin of life: The RNA world. *Nature* 319:618.

Gilbert, W., Marchionni, M., and McKnight, G. 1986. On the antiquity of introns. *Cell* 46: 151–53.

Ginoza, W., Hoelle, C. J., Vessey, K. B., and Carmack, C. 1964. Mechanisms of inactivation of single-stranded virus nucleic acids by heat. *Nature* 203:606–9.

Glansdorff, N. 1999. On the origin of operons and their possible role in evolution toward thermophily. *Journal of Molecular Evolution* 49:432–38.

Glansdorff, N., Xu, Y., and Labedan, B. 2008. The last universal common ancestor: Emergence, constitution and genetic legacy of an elusive forerunner. *Biology Direct* 3:29. doi:10.1186/1745 -6150-3-29.

Gogarten, J. P., Kibak, H., Dittrich, P., Taiz, L., Bowman, E. J., Bowman, B. J., et al. 1989. Evolution of the vacuolar H+-ATPase: Implications for the origin of eukaryotes. *Proceedings of the National Academy of Sciences USA* 86:6661–65.

Gogarten-Boekels, M., Hilario, E., and Gogarten, J. P. 1995. The effects of heavy meteorite bombardment on the early evolution: The emergence of the three domains of life. *Origins of Life and Evolution of Biospheres* 25:251–64.

Gold, T. 1992. The deep, hot biosphere. *Proceedings of the National Academy of Sciences USA* 89:6045–49.

Gomes, R., Levison, H. F., Tsiganis, K., and Morbidelli, A. 2005. Origin of the cataclysmic Late Heavy Bombardment period of the terrestrial planets. *Nature* 435:466–69.

Gorlas, A., Koonin, E. V., Bienvenu, N., Prieur, D., and Geslin, C. 2012. TPV1, the first virus isolated from the hyperthermophilic genus Thermococcus. *Environmental Microbiology* 14: 503–16.

Gorlas, A., Marguet, E., Gill, S., Geslin, C., Guigner, J. M., Guyot, F., et al. 2015. Sulfur vesicles from Thermococcales: A possible role in sulfur detoxifying mechanisms. *Biochimie* 118:356–64.

Gribaldo, S., and Brochier, C. 2009. Phylogeny of prokaryotes: Does it exist and why should we care? *Research in Microbiology* 160:513–21.

Gribaldo, S., Poole, A. M., Daubin, V., Forterre, P., and Brochier-Armanet, C. 2010. The origin of eukaryotes and their relationship with the Archaea: Are we at a phylogenomic impasse? *Nature Reviews Microbiology* 8:743–52.

Grosshans, C. A., and Cech, T. R. 1989. Metal ion requirements for sequence-specific endoribonuclease activity of the Tetrahymena ribozyme. *Biochemistry* 28:6888–94.

Groussin, M., and Gouy, M. 2011. Adaptation to environmental temperature is a major determinant of molecular evolutionary rates in archaea. *Molecular Biology and Evolution* 28: 2661–74.

Guerrier-Takada, C., Gardiner, K., Marsh, T., Pace, N., and Altman, S. 1983. The RNA moiety of ribonuclease P is the catalytic subunit of the enzyme. *Cell* 35:849–57.

Guipaud, O., Marguet, E., Noll, K. M., Bouchier de la Tour, C., and Forterre, P. 1997. Both DNA gyrase and reverse gyrase are present in the hyperthermophilic bacterium Thermotoga maritima. *Proceedings of the National Academy of Sciences USA* 94:10606–11.

Guy, L., and Ettema, T. J. G. 2011. The archaeal "TACK" superphylum and the origin of eukaryotes. *Trends in Microbiology* 19:580–87.

Häring, M., Rachel, R., Peng, X., Garrett, R. A., and Prangishvili, D. 2005b. Viral diversity in hot springs of Pozzuoli, Italy, and characterization of a unique archaeal virus, Acidianus bottle-shaped virus, from a new family, the Ampullaviridae. *Journal of Virology* 79:9904–11.

Häring, M., Vestergaard, G., Rachel, R., Chen, L., Garrett, R. A., and Prangishvili, D. 2005a. Virology: Independent virus development outside a host. *Nature* 436:1101–2.

Harold, F. M., and Van Brunt, J. 1977. Circulation of H+ and K+ across the plasma membrane is not obligatory for bacterial growth. *Science* 197:372–73.

Hartung, F., and Puchta, H. 2001. Molecular characterization of homologues of both subunits A (SPO11) and B of the archaebacterial topoisomerase 6 in plants. *Gene* 271:81–86.

Häusler, T. 2007. *Viruses vs. Superbugs.* Palgrave Macmillan.

Hennig, W. 1966. Phylogenetic systematics. *Annual Review of Entomology* 10:97–116.

Hethke, C., Bergerat, A., Hausner, W., Forterre, P., and Thomm, M. 1999. Cell-free transcription at 95 degrees: Thermostability of transcriptional components and DNA topology requirements of Pyrococcus transcription. *Genetics* 152:1325–33.

Huber, C., and Wächtershäuser, G. 2006. Alpha-hydroxy and alpha-amino acids under possible Hadean, volcanic origin-of-life conditions. *Science* 314:630–32.

Huber, R., Langworthy, T. A., König, H., Thomm, M., Woese, C. R., Sleytr, U. B., et al. 1986. Thermotoga maritima sp. Nov. Represents a new genus of unique extremely thermophilic eubacteria growing up to 90°C. *Archives of Microbiology* 144:324–33.

Huber, R., Stotters, P., Cheminee, J. L., Richnow, H. H., and Stetter, K. O. 1990. Hyperthermophilic archaebacteria within the crater and open-sea plume of erupting Macdonald Seamount. *Nature* 345:179–82.

Huber, R., Wilharm, T., Huber, D., Trincone, A., Burggraf, S., Konig, H., et al. 1992. Aquifex pyrophilus gen. nov. sp. nov., represents a novel group of marine hyperthermophilic hydrogen-oxidizing bacteria. *Systematic and Applied Microbiology* 15:340–51.

Huet, J., Schnabel, R., Sentenac, A., and Zillig, W. 1983. Archaebacteria and eukaryotes possess DNA-dependent RNA polymerases of a common type. *EMBO Journal* 2:1291–94.

Irimia, M., and Roy, S. W. 2014. Origin of spliceosomal introns and alternative splicing. *Cold Spring Harbor Perspectives in Biology* 6(6).

Iwabe, N., Kuma, K., Hasegawa, M., Osawa, S., and Miyata, T. 1989. Evolutionary relationship of archaebacteria, eubacteria, and eukaryotes inferred from phylogenetic trees of duplicated genes. *Proceedings of the National Academy of Sciences USA* 86:9355–59.

Izquierdo-Useros, N., Puertas, M. C., Borràs, F. E., Blanco, J., and Martinez-Picado, J. 2011. Exosomes and retroviruses: The chicken or the egg? *Cellular Microbiology* 13:10–17.

Jacob, F., and Monod, J. 1961. Genetic regulatory mechanisms in the synthesis of proteins. *Journal of Molecular Biology* 3:318–56.

Janekovic, S., Wunderl, I., Zillig, W., Gierl, A., and Neumann, H. 1983. TTV1, TTV2 and TTV3: A family of viruses of the extremely thermophilic, anaerobic, sulfur reducing archaebacterium Thermoproteus tenax. *Molecular and General Genetics* 192:39–45.

Jolivet, E., L'Haridon, S., Corre, E., Forterre, P., and Prieur, D. 2003. Thermococcus gammatolerans sp. nov., a hyperthermophilic archaeon from a deep-sea hydrothermal vent that resists ionizing radiation. *International Journal of Systematic and Evolutionary Microbiology* 53:847–51.

Jørgensen, S. L., Thorseth, I. H., Pedersen, R. B., Baumberger, T., and Schleper, C. 2013. Quantitative and phylogenetic study of the Deep Sea Archaeal Group in sediments of the Arctic mid-ocean spreading ridge. *Frontiers in Microbiology* 4:299. doi:10.3389/fmicb.2013.00299.

Joyce, G. F. 2002. The antiquity of RNA-based evolution. *Nature* 418:214–21.

Kandler, O., and König, H. 1978. Chemical composition of the peptidoglycan-free cell walls of methanogenic bacteria. *Archives of Microbiology* 118:141–52.

Karner, M. B., DeLong, E. F., and Karl, D. M. 2001. Archaeal dominance in the mesopelagic zone of the Pacific Ocean. *Nature* 409:507–10.

Kashefi, K. 2004. Response to Cowan: The upper temperature for life: Where do we draw the line? *Trends in Microbiology* 12:60–62.

Kashefi, K., and Lovley, D. R. 2003. Extending the upper temperature limit for life. *Science* 301:934.

Keller, M., Blöchl, E., Wächtershäuser, G., and Stetter, K. O. 1994. Formation of amide bonds without a condensation agent and implications for origin of life. *Nature* 368:836–38.

Kikuchi, A., and Asai, K. 1984. Reverse gyrase: A topoisomerase which introduces positive superhelical turns into DNA. *Nature* 309:677–81.

Klein, D. J., Moore, P. B., and Steitz, T. A. 2004. The contribution of metal ions to the structural stability of the large ribosomal subunit. *RNA* 10:1366–79.

Klein, R. A., Hazlewood, G. P., Kemp, P., and Dawson, R. M. 1979. A new series of long-chain dicarboxylic acids with vicinal dimethyl branching found as major components of the lipids of Butyrivibrio spp. *Biochemistry Journal* 183:691–700.

Koga, Y. 2012. Thermal adaptation of the archaeal and bacterial lipid membranes. *Archaea* ID 789652.

Konings, W. N. 2006. Microbial transport: Adaptations to natural environments. *Antonie Van Leeuwenhoek* 90:325–42.

Konings, W. N., Albers, S. V., Koning, S., and Driessen, A. J. 2002. The cell membrane plays a crucial role in survival of bacteria and archaea in extreme environments. *Antonie Van Leeuwenhoek* 81:61–72.

Könneke, M., Bernhard, A. E., de la Torre, J. R., Walker, C. B., Waterbury, J. B., and Stahl, D. A. 2005. Isolation of an autotrophic ammonia-oxidizing marine archaeon. *Nature* 437:543–46.

Koonin, E. V. 2003. Comparative genomics, minimal gene-sets and the last universal common ancestor. *Nature Reviews Microbiology* 1:127–36.

———. 2010a. The incredible expanding ancestor of eukaryotes. *Cell* 140:606–8.

———. 2010b. The origin and early evolution of eukaryotes in the light of phylogenomics. *Genome Biology* 11(5):209.

Koonin, E. V., Dolja, V. V., and Krupovic, M. 2015. Origins and evolution of viruses of eukaryotes: The ultimate modularity. *Virology* 479–80:2–25.

Koonin, E. V., and Krupovic, M. 2015. Evolution of adaptive immunity from transposable elements combined with innate immune systems. *Nature Reviews Genetics* 16:184–92.

Koonin, E. V., and Martin, W. 2005. On the origin of genomes and cells within inorganic compartments. *Trends in Genetics* 21(12):647–54.

Koonin, E. V., and Yutin, N. 2014. The dispersed archaeal eukaryome and the complex archaeal ancestor of eukaryotes. *Cold Spring Harbour Perspective in Biology* 6(4):a016188.

Kopylov, V. M., Bonch-Osmolovkaia, E. A., Svetlichny, V. A., Miroshnichenko, M. L., and Skobkin, V. S. 1993. Gamma-resistance and UV-sensitivity of extremely thermophilic archaebacteria and eubacteria. *Mikrobiolgiya* 62:90–95.

Kornberg, A. 1991. *For the Love of Enzymes.* Harvard University Press.

Kowalak, J. A., Dalluge, J. J., McCloskey, J. A., and Stetter, K. O. 1994. The role of posttranscriptional modification in stabilization of transfer RNA from hyperthermophiles. *Biochemistry* 33:7869–76.

Kruger, K., Grabowski, P. J., Zaug, A. J., Sands, J., Gottschling, D. E., and Cech, T. R. 1982. Self-splicing RNA: Autoexcision and autocyclization of the ribosomal RNA intervening sequence of Tetrahymena. *Cell* 31:147–57.

Krupovic, M., and Bamford, D. H. 2010. Order to the viral universe. *Journal of Virology* 84:12476–79.

Krupovic, M., Makarova, K. S., Forterre, P., Prangishvili, D., and Koonin, E. V. 2014a. Caspo sons: A new superfamily of self-synthesizing DNA transposons at the origin of prokaryotic CRISPR-Cas immunity. *BMC Biology* 12:36.

Krupovic, M., Quemin, E. R., Bamford, D. H., Forterre, P., and Prangishvili, D. 2014b. Unification of the globally distributed spindle-shaped viruses of the Archaea. *Journal of Virology* 88:2354–58.

Kuhn, P., Lesley, S. A., Mathews I. I., Canaves, J. M., Brinen, L. S., Dai, X., et al. 2002. A Crystal structure of thy1, a thymidylate synthase complementing protein from Thermotoga maritima at 2.25 A resolution. *Proteins* 49:142–45.

Kulp, A., and Kuehn, M. J. 2010. Biological functions and biogenesis of secreted bacterial outer membrane vesicles. *Annual Review of Microbiology* 64:163–84.

Kurland, C. G., Collins, L. J., and Penny, D. 2006. Genomics and the irreducible nature of eukaryote cells. *Science* 312:1011–14.

Kurr, M., Huber, R., Konig, H., Jannasch, H. W., Fricke, H., Trincone, A., et al. 1991. Methanopyrus kandleri, gen. and sp. nov. represents a novel group of hyperthermophilic methanogens, growing at 110°C. *Archives of Microbiology* 156:239–47.

Lagorce, A., Fourçans, A., Dutertre, M., Bouyssiere, B., Zivanovic, Y., and Confalonieri, F. 2012. Genome-wide transcriptional response of the archaeon Thermococcus gammatolerans to cadmium. *PLoS One* 7(7):e41935.

Lai, M. M. 2005. RNA replication without RNA-dependent RNA polymerase: Surprises from hepatitis delta virus. *Journal of Virology* 79:7951–58.

Lake, J. A., Henderson, E., Oakes, M., and Clark, M. W. 1984. Eocytes: A new ribosome structure indicates a kingdom with a close relationship to eukaryotes. *Proceedings of the National Academy of Sciences USA* 81:3786–90.

Lane, C. E., Van Den Heuvel, K., Kozera, C., Curtis, B. A., Parsons, B. J., et al. 2007. Nucleomorph genome of Hemiselmis andersenii reveals complete intron loss and compaction as a driver of protein structure and function. *Proceedings of the National Academy of Sciences USA* 104:19908–13.

Lane, N., and Martin, W. 2010. The energetics of genome complexity. *Nature* 467:929–34.

Langworthy, T. A., Smith, P. F., and Mayberry, W. R. 1972. Lipids of Thermoplasma acidophilum. *Journal of Bacteriology* 112:1193–200.

———. 1974. Long-chain glycerol diether and polyol dialkyl glycerol triether lipids of Sulfolobus acidocaldarius. *Journal of Bacteriology* 119:106–16.

La Scola, B., Desnues, C., Pagnier, I., Robert, C., Barrassi, L., Fournous, G., et al. 2008. The virophage as a unique parasite of the giant mimimivirus. *Nature* 455:100–104.

Lazcano, A., Guerrero, R., Margulis, L., and Oró, J. 1988. The evolutionary transition from RNA to DNA in early cells. *Journal of Molecular Evolution* 27:283–90.

Lecompte, O., Ripp, R., Thierry, J. C., Moras, D., and Poch, O. 2002. Comparative analysis of ribosomal proteins in complete genomes: An example of reductive evolution at the domain scale. *Nucleic Acids Research* 30:5382–90.

Leipe, D. D., Aravind, L., and Koonin, E. V. 1999. Did DNA replication evolve twice independently? *Nucleic Acids Research* 27:3389–401.

Lepage, E., Marguet, E., Geslin, C., Matte-Tailliez, O., Zillig, W., Forterre, P., et al. 2004. Molecular diversity of new Thermococcales isolates from a single area of hydrothermal deep-sea vents as revealed by randomly amplified polymorphic DNA fingerprinting and 16S rRNA gene sequence analysis. *Applied Environmental Microbiology* 70:1277–86.

Lindahl, T. 1967. Irreversible heat inactivation of transfer ribonucleic acids. *Journal of Biological Chemistry* 242:1970–73.

Lindås, A. C., Karlsson, E. A., Lindgren, M. T., Ettema, T. J., and Bernander, R. 2008. A unique cell division machinery in the Archaea. *Proceedings of the National Academy of Sciences USA* 105:18942–46.

Liu, H., Rudolf, J., Johnson, K. A., McMahon, S. A., Oke, M., Carter, L., et al. 2008. Structure of the DNA repair helicase XPD. *Cell* 133:801–12.

Liu, L. F., Liu, C. C., and Alberts, B. M. 1979. T4 DNA topoisomerase: A new ATP-dependent enzyme essential for initiation of T4 bacteriophage DNA replication. *Nature* 281:456–61.

Llanos, J., Capasso, C., Parisi, E., Prieur, D., and Jeanthon, C. 2000. Susceptibility to heavy metals and cadmium accumulation in aerobic and anaerobic thermophilic microorganisms isolated from deep-sea hydrothermal vents. *Current Microbiology* 41:201–5.

Lombard, J., López-García, P., and Moreira, D. 2012. The early evolution of lipid membranes and the three domains of life. *Nature Reviews Microbiology* 10:507–15.

López-García, P., Forterre, P., van der Oost, J., and Erauso, G. 2000. Plasmid pGS5 from the hyperthermophilic archaeon Archaeoglobus profundus is negatively supercoiled. *Journal of Bacteriology* 182:4998–5000.

López-García, P., and Moreira, D. 2006. Selective forces for the origin of the eukaryotic nucleus. *BioEssays* 28:525–33.

Lwoff, A. 1953. Lysogeny. *Bacteriological Reviews* 17:269–337.

Maaty W. S., Ortmann A. C., Dlakić M., Schulstad K., Hilmer J. K., Liepold L., et al. 2006. Characterization of the archaeal thermophile Sulfolobus turreted icosahedral virus validates an evolutionary link among double-stranded DNA viruses from all domains of life. *Journal of Virology* 80:7625–35.

Magrum, L. J., Luehrsen, K. R., and Woese, C. R. 1978. Are extreme halophiles actually "bacteria"? *Journal of Molecular Evolution* 11:1–8.

Marc, F., Sandman, K., Lurz, R., and Reeve, J. N. 2002. Archaeal histone tetramerization determines DNA affinity and the direction of DNA supercoiling. *Journal of Biological Chemistry* 277:30879–86.

Marguet, E., and Forterre, P. 1994. DNA stability at temperatures typical for hyperthermophiles. *Nucleic Acids Research* 22:1681–86.

———. 1998. Protection of DNA by salts against thermodegradation at temperatures typical for hyperthermophiles. *Extremophiles* 2:115–22.

Marguet, E., Gaudin, M., Gauliard, E., Fourquaux, I., le Blond du Plouy, S., Matsui, I., et al. 2013. Membrane vesicles, nanopods and/or nanotubes produced by hyperthermophilic archaea of the genus Thermococcus. *Biochemical Society Transactions* 41:436–42.

Marteinsson, V. T., Bjornsdottir, S. H., Bienvenu, N., Kristjansson, J. K., and Birrien, J. L. 2010. Rhodothermus profundi sp. nov., a thermophilic bacterium isolated from a deep-sea hydrothermal vent in the Pacific Ocean. *International Journal of Systematic and Evolutionary Microbiology* 60:2729–34.

Martijn, J., and Ettema, T. J. 2013. From archaeon to eukaryote: The evolutionary dark ages of the eukaryotic cell. *Biochemical Society Transactions* 41:451–57.

Martin, A., Yeats, S., Janekovic, D., Reiter, W. D., Aicher, W., and Zillig, W. 1984. SAV 1, a temperate u.v.-inducible DNA virus-like particle from the archaebacterium Sulfolobus acidocaldarius isolate B12. *EMBO Journal* 3:2165–68.

Martin, W., Baross, J., Kelley, D., and Russell, M. J. 2008. Hydrothermal vents and the origin of life. *Nature Reviews Microbiology* 6:805–14.

Martin, W., and Koonin, E. V. 2006. Introns and the origin of nucleus-cytosol compartmentalization. *Nature* 440:41–45.

Martin, W., and Russell, M. J. 2007. On the origin of biochemistry at an alkaline hydrothermal vent. *Philosophical Transactions of the Royal Society of London. Series B: Biological Sciences* 3621887–925.

Mashburn, L. M., and Whiteley, M. 2005. Membrane vesicles traffic signals and facilitate group activities in a prokaryote. *Nature* 437:422–25.

Matsunaga, F., Forterre, P., Ishino, Y., and Myllykallio, H. 2001. *In vivo* interactions of archaeal Cdc6/Orc1 and minichromosome maintenance proteins with the replication origin. *Proceedings of the National Academy of Sciences USA* 98:11152–57.

McCloskey, J. A., Graham, D. E., Zhou, S., Crain, P. F., Ibba, M., Konisky, J., et al. 2001. Posttranscriptional modification in archaeal tRNAs: Identities and phylogenetic relations of nucleotides from mesophilic and hyperthermophilic Methanococcales. *Nucleic Acids Research* 29:4699–706.

McInerney, J. O., O'Connell, M. J., and Pisani, D. 2014. The hybrid nature of the Eukaryota and a consilient view of life on Earth. *Nature Reviews Microbiology* 12:449–55.

Meckes, D. G. Jr. 2015. Exosomal communication goes viral. *Journal of Virology* 89:5200–3.

Melo, S. A., Luecke, L. B., Kahlert, C., Fernandez, A. F., Gammon, S. T., Kaye, J., et al. 2015. Glypican-1 identifies cancer exosomes and detects early pancreatic cancer. *Nature* 523:177–82.

Miller, S. L. 1953. A production of amino acids under possible primitive earth conditions. *Science* 117:528–29.

Miller, S. L., and Lazcano, A. 1995. The origin of life: Did it occur at high temperatures? *Journal of Molecular Evolution* 41:689–92.

Mitchell, P. 1961. Coupling of Phosphorylation to Electron and Hydrogen Transfer by a Chemi-Osmotic type of Mechanism. *Nature* 191:144–48.

Moreira, D., and López-García, P. 2009. Ten reasons to exclude viruses from the tree of life. *Nature Reviews Microbiology* 7:306–11.

Mulkidjanian, A. Y., Bychkov, A. Y., Dibrova, D. V., Galperin, M. Y., and Koonin, E. V. 2012. Origin of first cells at terrestrial, anoxic geothermal fields. *Proceedings of the National Academy of Sciences USA* 109(14):E821–30.

Mulkidjanian, A. Y., Makarova, K. S., Galperin, M. Y., and Koonin, E. V. 2007. Inventing the dynamo machine: The evolution of the F-type and V-type ATPases. *Nature Reviews Microbiology* 5:892–99.

Mushegian, A. R., and Koonin, E. V. 1996. A minimal gene set for cellular life derived by comparison of complete bacterial genomes. *Proceedings of the National Academy of Sciences USA* 93:10268–73.

Myllykallio, H., Lipowski, G., Leduc, D., Filee, J., Forterre, P., and Liebl, U. 2002. An alternative flavin-dependent mechanism for thymidylate synthesis. *Science* 297:105–7.

Myllykallio, H., Lopez, P., López-García, P., Heilig, R., Saurin, W., Zivanovic, Y., Philippe, H., et al. 2000. Bacterial mode of replication with eukaryotic-like machinery in a hyperthermophilic archaeon. *Science* 288:2212–15.

Nakasu, S., and Kikuchi, A. 1985. Reverse gyrase; ATP-dependent type I topoisomerase from Sulfolobus. *EMBO Journal* 4:2705–10.

Nelson, K. E., Clayton, R. A., Gill, S. R., Gwinn, M. L., Dodson, R. J., Haft, D. H., et al. 1999. Evidence for lateral gene transfer between Archaea and bacteria from genome sequence of Thermotoga maritima. *Nature* 399:323–29.

Nercessian, O., Reysenbach, A. L., Prieur, D., and Jeanthon, C. 2003. Archaeal diversity associated with in situ samplers deployed on hydrothermal vents on the East Pacific Rise (13 degrees N). *Environmental Microbiology* 5:492–502.

Noon, K. R., Guymon, R., Crain, P. F., McCloskey, J. A., Thomm, M., Lim, J., et al. 2003. Influence of temperature on tRNA modification in archaea: Methanococcoides burtonii (optimum growth temperature [Topt], 23 degrees C) and Stetteria hydrogenophila (Topt, 95 degrees C). *Journal of Bacteriology* 185:5483–90.

Nunoura, T., Takaki, Y., Kakuta, J., Nishi, S., Sugahara, J., Kazama, H., et al. 2011. Insights into the evolution of Archaea and eukaryotic protein modifier systems revealed by the genome of a novel archaeal group. *Nucleic Acids Research* 39:3204–23.

Oberto, J., Gaudin, M., Cossu, M., Gorlas, A., Slesarev, A., Marguet, E., et al. 2014. Genome sequence of a hyperthermophilic archaeon, Thermococcus nautili 30–1, that produces viral vesicles. *Genome Announcements* 2. pii: e00243–14. doi:10.1128/genomeA.00243-14.

Olby, R. 1994. *The Path to the Double Helix: The Discovery of DNA*. University of Washington Press. 1974 & rev. 1994.

Olsen G. J., and Woese, C. R. 1997. Archaeal genomics: An overview. *Cell* 89(7):991–94. Review.

Oparin, A. I. 1959. *The Origin of Life on the Earth*. Pergamon Press.

Orgel, L. E. 1968. Evolution of the genetic apparatus. *Journal of Molecular Biology* 38: 381–93.

Pace, N. R. 2009. Problems with "procaryote." *Journal of Bacteriology* 191:2008–10.

Pace, N. R., Olsen, G. J., and Woese, C. R. 1986. Ribosomal RNA phylogeny and the primary lines of evolutionary descent. *Cell* 45:325–26.

Peak, M. J., Robb, F. T., and Peak, J. G. 1995. Extreme resistance to thermally induced DNA backbone breaks in the hyperthermophilic archaeon Pyrococcus furiosus. *Journal of Bacteriology* 177:6316–18.

Pedersen, S., and Reeh, S. 1978. Functional mRNA half lives in E. coli. *Molecular and General Genetics* 166:329–36.

Penny, D., Hoeppner, M. P., Poole, A. M., and Jeffares, D. C. 2009. An overview of the introns-first theory. *Journal of Molecular Evolution* 69:527–40.

Peretó J., López-García, P., and Moreira, D. 2004. Ancestral lipid biosynthesis and early membrane evolution. *Trends in Biochemical Sciences* 29:469–77.

Perrochia, L., Crozat, E., Hecker, A., Zhang, W., Bareille, J., Collinet, B., et al. 2013. *In vitro* biosynthesis of a universal t6A tRNA modification in Archaea and Eukarya. *Nucleic Acids Research* 41:1953–64.

Pester, M., Schleper, C., and Wagner, M. 2011. The Thaumarchaeota: An emerging view of their phylogeny and ecophysiology. *Current Opinion in Microbiology* 14:300–306.

Pietilä M. K., Laurinmäki, P., Russell, D. A., Ko, C. C., Jacobs-Sera, D., Hendrix, R. W., et al. 2013. Structure of the archaeal head-tailed virus HSTV-1 completes the HK97 fold story. *Proceedings of the National Academy of Sciences USA* 110:10604–9.

Pino, S., Sponer, J. E., Costanzo, G., Saladino, R., and Mauro, E. D. 2015. From formamide to RNA, the path is tenuous but continuous. *Life* 5:372–84.

Pommier, Y., Leo, E., Zhang, H., and Marchand, C. 2010. DNA topoisomerases and their poisoning by anticancer and antibacterial drugs. *Chemical Biology* 17:421–33.

Poole, A., Jeffares, D., and Penny, D. 1999. Early evolution: Prokaryotes, the new kids on the block. *Bioessays* 21:880–89.

Poole, A. M. 2009. Horizontal gene transfer and the earliest stages of the evolution of life. *Research in Microbiology* 160:473–80.

Poole, A. M., and Logan, D. T. 2005. Modern mRNA proofreading and repair: Clues that the last universal common ancestor possessed an RNA genome? *Molecular Biology and Evolution* 22:1444–55.

Prangishvili, D. 2013. The wonderful world of archaeal viruses. *Annual Review of Microbiology* 67:565–85.

Prangishvili, D., Arnold, H. P., Götz, D., Ziese, U., Holz, I., Kristjansson, J. K., et al. 1999. A novel virus family, the Rudiviridae: Structure, virus-host interactions and genome variability of the sulfolobus viruses SIRV1 and SIRV2. *Genetics* 152:1387–96.

Prangishvili, D., Garrett, R. A., and Koonin, E. V. 2006. Evolutionary genomics of archaeal viruses: Unique viral genomes in the third domain of life. *Virus Research* 117:52–67.

Prangishvili, D., and Krupovic, M. 2012. A new proposed taxon for double-stranded DNA viruses, the order "Ligamenvirales." *Archives of Virology* 157:791–95.

Prangishvili, D., Zillig, W., Gierl, A., Biesert, L., and Holz, I. 1982. DNA-dependent RNA polymerase of thermoacidophilic archaebacteria. *European Journal of Biochemistry* 122:471–77.

Probst, A. J., Auerbach, A. K., and Moissl-Eichinger, C. 2013. Archaea on human skin. *PLoS One* 8:e65388. doi:10.1371/journal.pone.0065388.

Quax, T. E., Lucas, S., Reimann, J., Pehau-Arnaudet, G., Prevost, M. C., Forterre, P., et al. 2011. Simple and elegant design of a virion egress structure in Archaea. *Proceedings of the National Academy of Sciences USA* 108:3354–59.

Quemin, E. R., Lucas, S., Daum, B., Quax, T. E., Kühlbrandt, W., Forterre, P., et al. 2013. First insights into the entry process of hyperthermophilic archaeal viruses. *Journal of Virology* 87:13379–85.

Raoult, D., Audic, S., Robert, C., Abergel, C., Renesto, P., Ogata, H., et al. 2004. The 1.2-megabase genome sequence of Mimivirus. *Science* 306:1344–50.

Raoult, D., and Forterre, P. 2008. Redefining viruses: Lessons from Mimivirus. *Nature Reviews Microbiology* 2008 6:315–19.

Ravaux, J., Hamel, G., Zbinden, M., Tasiemski, A. A., Boutet, I., Léger, N., et al. 2013. Thermal limit for metazoan life in question: *In vivo* heat tolerance of the Pompeii worm. *PLoS One* 8(5):e64074. doi:10.1371/journal.pone.0064074.

Raymann, K., Brochier-Armanet, C., and Gribaldo, S. 2015. The two-domain tree of life is linked to a new root for the Archaea. *Proceedings of the National Academy of Sciences USA* 112:6670–75.

Rice, D. W., Yip, K. S., Stillman, T. J., Britton, K. L., Fuentes, A., Connerton, I., et al. 1996. Insights into the molecular basis of thermal stability from the structure determination of Pyrococcus furiosus glutamate dehydrogenase. *FEMS Microbiology Reviews* 18:105–17.

Rinke, C., Schwientek, P., Sczyrba, A., Ivanova, N. N., Anderson, I. J., Cheng, J. F., et al. 2013. Insights into the phylogeny and coding potential of microbial dark matter. *Nature* 499:431–37.

Robert, F., and Chaussidon, M. 2006. A palaeotemperature curve for the Precambrian oceans based on silicon isotopes in cherts. *Nature* 443:969–72.

Rodríguez-Trelles, F., Tarrío, R., and Ayala, F. J. 2006. Origins and evolution of spliceosomal introns. *Annual Review of Genetics* 40:47–76.

Rogozin, I. B., Carmel, L., Csuros, M. and Koonin, E. V. 2012. Origin and evolution of spliceosomal introns. *Biology Direct* 7,11. doi:10.1186/1745-6150-7-11.

Rouillon, C., and White, M. F. 2011. The evolution and mechanisms of nucleotide excision repair proteins. *Research in Microbiology* 162:19–26.

Russell, M. J., Barge, L. M., Bhartia, R., Bocanegra, D., Bracher, P. J., Branscomb, E., et al. 2014. The drive to life on wet and icy worlds. *Astrobiology* 14:308–43.

Sabath, N., Wagner, A., and Karlin, D. 2012. Evolution of viral proteins originated *de novo* by overprinting. *Molecular Biology and Evolution* 29:3767–80.

Saiki, R. K., Gelfand, D. H., Stoffel, S., Scharf, S. J., Higuchi, R., Horn, G. T., et al. 1988. Primer-directed enzymatic amplification of DNA with a thermostable DNA polymerase. *Science* 239:487–91.

Samson, R. Y., Obita, T., Freund, S. M., Williams, R. L., and Bell, S. D. 2008. A role for the ESCRT system in cell division in archaea. *Science* 322:1710–13.

Sapp, J., 2009. *The New Foundations of Evolution: On the Tree of Life.* Oxford University Press.

Schleper, C., Kubo, K., and Zillig, W. 1992. The particle SSV1 from the extremely thermophilic archaeon Sulfolobus is a virus: Demonstration of infectivity and of transfection with viral DNA. *Proceedings of the National Academy of Sciences USA* 89:7645–49.

Schmidt, T. M., DeLong, E. F., and Pace, N. R. 1991. Analysis of a marine picoplankton community by 16S rRNA gene cloning and sequencing. *Journal of Bacteriology* 173:4371–78.

Sehgal, S. N., Kates, M., and Gibbons, N. E. 1962. Lipids of Halobacterium cutirubrum. *Canadian Journal of Biochemistry and Physiology* 40:69–81.

Setchell, W. A. 1903. The upper temperature limit of life. *Science* 17:934–37.

She, Q., Singh, R. K., Confalonieri, F., Zivanovic, Y., Allard, G., Awayez, M. J., et al. 2001. The complete genome of the crenarchaeon Sulfolobus solfataricus P2. *Proceedings of the National Academy of Sciences USA* 98:7835–40.

Shetty, A., Chen, S., Tocheva, E. I., Jensen, G. J., and Hickey, W. J. 2011. Nanopods: A new bacterial structure and mechanism for deployment of outer membrane vesicles. *PLoS One* (6):e20725 doi:10.1371/journal.pone.0020725.

Shin, D. S., Pratt, A. J., and Tainer, J. A. 2014. Archaeal genome guardians give insights into eukaryotic DNA replication and damage response proteins. *Archaea* 20;2014:206735. doi: 10.1155/2014/206735. eCollection 2014.

Skouloubris, S., Djaout, K., Lamarre, I., Lambry, J. C., Anger, K., Briffotaux, J., et al. 2015. Targeting of Helicobacter pylori thymidylate synthase ThyX by non-mitotoxic hydroxynaphthoquinones. *Open Biology* 5(6):150015. doi:10.1098/rsob.150015.

Soler, N., Krupovic, M., Marguet, E., and Forterre, P. 2015. Membrane vesicles in natural environments: A major challenge in viral ecology. *ISME Journal: Multidisciplinary Journal of Microbial Ecology* 9:793–96.

Soler, N., Marguet, E., Cortez, D., Desnoues, N., Keller, J., van Tilbeurgh, H., et al. 2010. Two novel families of plasmids from hyperthermophilic archaea encoding new families of replication proteins. *Nucleic Acids Research* 38:5088–104.

Soler, N., Marguet, E., Verbavatz J. M., and Forterre, P. 2008. Virus-like vesicles and extracellular DNA produced by hyperthermophilic archaea of the order Thermococcales. *Research in Microbiology* 159:390–99.

Spang, A., Poehlein, A., Offre, P., Zumbrägel, S., Haider, S., Rychlik, N., et al. 2012. The genome of the ammonia-oxidizing Candidatus Nitrososphaera gargensis: Insights into metabolic versatility and environmental adaptations. *Environmental Microbiology* 14:3122–45.

Spang, A., Saw, J. H., Jørgensen, S. L., Zaremba-Niedzwiedzka, K., Martijn, J., Lind, A. E., et al. 2015. Complex archaea that bridge the gap between prokaryotes and eukaryotes. *Nature* 521:173–79.

Sprott, G. D., Meloche, M., and Richards, J. C. 1991. Proportions of diether, macrocyclic diether, and tetraether lipids in Methanococcus jannaschii grown at different temperatures. *Journal of Bacteriology* 173:3907–10.

Sternberg, S. H., and Doudna, J. A. 2015. Expanding the biologist's toolkit with CRISPR-Cas9. *Molecular Cell* 58:568–74.

Stetter, K. O. 1982. Ultrathin mycelia-forming organisms from submarine volcanic areas having an optimum growth temperature of 105°C. *Nature* 300:258–60.

———. 1989. Extremely thermophilic chemolithoautotrophic archaebacteria. In *Autotrophic bacteria*. Eds. H. G. Schlegel and B. Bowen. Science Tech. Publishers and Springer Verlag, Berlin, 167–71.

———. 1996. Hyperthermophiles in the history of life. *Ciba Foundation Symposium* 202, 1–10.

———. 2013. A brief history of the discovery of hyperthermophilic life. *Biochemical Society Transactions* 41:416–20.

Stetter, K. O., Huber, R., Blöchl, E., Kurr, M., Eden, R.D., Flelder, M., et al. 1993. Hyperthermophilic archaea are thriving in deep North Sea and Alaskan oil reservoirs. *Nature* 365:743–45.

Stetter, K. O., König, H., and Stackebrandt, E. 1983. Pyrodictium gen. nov., a new genus of submarine disc-shaped sulphur reducing archaebacteria growing optimally at 105°C. *Systematic and Applied Microbiology* 4:535–51.

Stetter, K. O., Thomm, M., Winter, J., Wildgruber, G., Huber, H., Zillig, W., et al. 1981. Methanothermus fervidus, sp. nov., a novel extremely thermophilic methanogen isolated from an Icelandic hot spring. *Zentralblatt für Mikrobiologie und Hygiene: I. Abt. Originale C* 2:166–78.

Sugimoto-Shirasu, K., Stacey, N. J., Corsar, J., Roberts, K., and McCann, M. C. 2002. DNA topoisomerase VI is essential for endoreduplication in Arabidopsis. *Current Biology* 12:1782–86.

Takai, K., Nakamura, K., Toki, T., Tsunogai, U., Miyazaki, M., Miyazaki, J., et al. 2008. Cell proliferation at 122 degrees C and isotopically heavy CH4 production by a hyperthermophilic methanogen under high-pressure cultivation. *Proceedings of the National Academy of Sciences USA* 105:10949–54.

Temin, H. M. 1971. The protovirus hypothesis: Speculations on the significance of RNA-directed DNA synthesis for normal development and for carcinogenesis. *Journal of the National Cancer Institute* 46:3–7.

Théry, C. 2015. Cancer: Diagnosis by extracellular vesicles. *Nature* 523:161–62.

Thiaville, P. C., El Yacoubi, B., Perrochia, L., Hecker, A., Prigent, M., et al. 2014b. Cross kingdom functional conservation of the core universally conserved threonylcarbamoyladenosine tRNA synthesis enzymes. *Eukaryotic Cell* 13:1222–31.

Thiaville, P. C., Iwata-Reuyl, D., and de Crécy-Lagard, V. 2014a. Diversity of the biosynthesis pathway for threonylcarbamoyladenosine (t(6)A), a universal modification of tRNA. *RNA Biology* 11:1529–39.

Tornabene, T. G., and Langworthy, T. A. 1979. Diphytanyl and dibiphytanyl glycerol ether lipids of methanogenic archaebacteria. *Science* 203:51–53.

Trent, J. D., Chastain, R. A., Yayanos, A. A. 1984. Possible artefactual basis for apparent bacterial growth at 250 degrees C. *Nature* 307:737–40.

Unsworth, L. D., Van der Oost, J., and Koutsopoulos, S. 2007. Hyperthermophilic enzymes: Stability, activity and implementation strategies for high temperature applications. *Federation of the European Biochemical Societies Journal* 274:4044–56.

Valadij, H., Ekström, K., Bossios, A., Sjöstrand, M., Lee, J., and Lötvall, J. O. 2007. Exosome-mediated transfer of mRNAs and microRNAs is a novel mechanism of gene exchange between cells. *Nature Cell Biology* 9:654–59.

Valentine, D. L. 2007. Adaptations to energy stress dictate the ecology and evolution of the Archaea. *Nature Reviews Microbiology* 5:316–23.

Van de Vossenberg, J. L., Ubbink-Kok, T., Elferink, M. G., Driessen, A. J., and Konings, W. N. 1995. Ion permeability of the cytoplasmic membrane limits the maximum growth temperature of bacteria and archaea. *Molecular Microbiology* 18:925–32.

Wächtershäuser, G. 1988. Before enzymes and templates: Theory of surface metabolism. *Microbiological Reviews* 52:452–84.

Wan, L. C. K., Mao, D. Y. L., Neculai, D., Strecker, J., Chiovitti, D., Kurinov, I., et al. 2013. Reconstitution and characterization of eukaryotic N6-threonylcarbamoylation of tRNA using a minimal enzyme system. *Nucleic Acids Research* 2013; 41:6332–46.

Wang, J. C. 1971. Interaction between DNA and an Escherichia coli protein omega. *Journal of Molecular Biology* 55:523–33.

———. 2009. *Untangling the Double-Helix.* Cold Spring Harbor University Press.

Warren, R. A. 1980. Modified bases in bacteriophage DNAs. *Annual Review of Microbiology* 34:137–58.

Watson, J. D., and Crick, F. H. 1953a. Molecular structure of nucleic acids: A structure for deoxyribose nucleic acid. *Nature* 171:737–38.

———. 1953b. The structure of DNA. *Cold Spring Harbor Symposium on Quantitative Biology* 18:123–31.

White, R. H. 1984. Hydrolytic stability of biomolecules at high temperatures and its implication for life at 250 degrees C. *Nature*. 310:430–32.

Whitfield, J. 2006. Origins of DNA: Base invaders. *Nature* 439:130–31.

Wiegel, J., and Adams, W. W. M. 1998. *Thermophiles: The Keys to the Molecular Evolution and the Origin of Life?* CRC Press.

Williams, T. A., Foster, P. G., Cox, C. J., and Embley, T. M. 2013. An archaeal origin of eukaryotes supports only two primary domains of life. *Nature* 504:231–36.

Winans, S. C. 2005. Microbiology: Bacterial speech bubbles. *Nature* 437:330.

Woese, C. R. 1967. *The Genetic Code: The Molecular Basis for Genetic Expression*. Harper & Row, 186.

———. 1981. Archaebacteria. *Scientific American* 244:98–122.

———. 1987. Bacterial evolution. *Microbiological Reviews* 51:221–71.

———. 1998. The universal ancestor. *Proceedings of the National Academy of Sciences USA* 95:6854–59.

———. 2002. On the evolution of cells. *Proceedings of the National Academy of Sciences USA* 99:8742–47.

Woese, C. R., and Fox, G. E. 1977a. Phylogenetic structure of the prokaryotic domain: The primary kingdoms. *Proceedings of the National Academy of Sciences USA* 74:5088–90.

———. 1977b. The concept of cellular evolution. *Journal of Molecular Evolution* 10:1–6.

Woese, C. R., Kandler, O., and Wheelis, M. L. 1990. Towards a natural system of organisms: Proposal for the domains Archaea, Bacteria, and Eucarya. *Proceedings of the National Academy of Sciences USA* 87:4576–79.

Woese, C. R., Maniloff J., and Zablen L. B. 1980. Phylogenetic analysis of the mycoplasmas. *Proceedings of the National Academy of Sciences USA* 77:494–98.

Woese, C. R., and Olsen, G. J. 1986. Archaebacterial phylogeny: Perspectives on the urkingdoms. *Systematic and Applied Microbiology* 7:161–77.

Yates, D. 2007, October 16. Symposium marks 30th anniversary of discovery of third domain of life. *Illinois News Bureau*. Retrieved from https://news.illinois.edu/blog/view/6367/206509.

Yoon, H. J., Hossain, M. S., Held, M., Hou, H., Kehl, M., Tromas, A., et al. 2014. Lotus japonicus SUNERGOS1 encodes a predicted subunit A of a DNA topoisomerase VI that is required for nodule differentiation and accommodation of rhizobial infection. *Plant Journal* 78:811–21.

Zablen, L. B., Kissil, M. S., Woese, C. R., and Buetow, D. E. 1975. Phylogenetic origin of the chloroplast and prokaryotic nature of its ribosomal RNA. *Proceedings of the National Academy of Sciences USA* 72:2418–22.

Zhang, C., Tian, B., Li, S., Ao, X., Dalgaard, K., Gökce, S., et al. 2013. Genetic manipulation in Sulfolobus islandicus and functional analysis of DNA repair genes. *Biochemical Society Transactions* 41:405–10.

Zillig, W., Gierl, A., Schreiber, G., Wunderl, S., Janekovic, D., Stetter, K. O., et al. 1983a. The Archaebacterium Thermofilum pendens represents a novel genus of the thermophilic, anaerobic sulfur respiring Thermoproteales. *Systematic and Applied Microbiology* 4:79–87.

Zillig, W., Holz, I., Janekovic, D., Schäfer, W., and Reiter, W. D. 1983b. The Archaebacterium Thermococcus celer represents a novel genus within the thermophilic branch of the Archaebacteria. *Systematic and Applied Microbiology* 4:88–94.

Zillig, W., Stetter, K. O., and Janeković, D. 1979. DNA-dependent RNA polymerase from the archaebacterium Sulfolobus acidocaldarius. *European Journal of Biochemistry* 96:597–604.

Zillig W., Stetter, K. O., Prangishvili, D., Schäfer, W., Wunderl, S., Janekovic D., Holz, I., and Palm, P. 1982. Desulfurococcaceae, the second family of the extremely anaerobic, sulfur respiring Thermoproteales. *Zentralblatt für Bakteriologie Mikrobiologie und Hygiene: I. Abt. Originale* 3(2):304–17.

Zillig, W., Stetter, K. O., and Tobien, M. 1978. DNA-dependent RNA polymerase from Halobacterium halobium. *European Journal of Biochemistry* 91:193–99.

Zillig, W., Tu J., and Holz, I. 1981. Thermoproteales: A third order of thermoacidophilic archaebacteria. *Nature* 293:85–86.

Zimmer, C. 2006. Did DNA come from viruses? *Science* 312:870–72.

Zivanovic, Y., Armengaud, J., Lagorce, A., Leplat, C., Guérin, P., Dutertre, M., et al. 2009. Genome analysis and genome-wide proteomics of Thermococcus gammatolerans, the most radioresistant organism known amongst the Archaea. *Genome Biology* 10(6):R70.

Index

Page numbers in italics refer to figures.

Subjects